European Federation of Corrosio
NUMBER 64

T0251745

Recommended practice for corrosion management of pipelines in oil and gas production and transportation

Edited by
Bijan Kermani and Thierry Chevrot

EUROPEAN FEDERATION OF CORROSION
FÉDÉRATION EUROPÉENNE DE LA CORROSION
EUROPÄISCHE FÖDERATION KORROSION

Published for the European Federation of Corrosion
by CRC Press
on behalf of
The Institute of Materials, Minerals & Mining

CRC Press
Taylor & Francis Group
Boca Raton London New York

CRC Press is an imprint of the
Taylor & Francis Group, an informa business

The Institute of Materials,
Minerals and Mining

Published by Maney Publishing on behalf of the European Federation of Corrosion
and The Institute of Materials, Minerals & Mining

First published 2012 by Maney Publishing

Published 2017 by CRC Press
711 Third Avenue, New York, NY 10017, USA
2 Park Square, Milton Park, Abingdon, Oxon OX14 4RN

ISBN: 978-1-907-97533-2 (pbk)

ISSN 1354-5116

Contents

European Federation of Corrosion (EFC) publications: Series introduction

The EFC, founded in 1955, is a Federation of 33 societies with interests in corrosion and is based in twenty-six different countries throughout Europe and beyond. Its member societies represent the corrosion interests of more than 25,000 engineers, scientists and technicians. The Federation's aim is to advance the science of the corrosion and protection of materials by promoting cooperation in Europe and collaboration internationally. Asides from national and international corrosion societies, universities, research centres and companies can also become Affiliate Members of the EFC.

The administration of the Federation is in the hands of the Board of Administrators, chaired by the EFC President, and the scientific and technical affairs are the responsibility of the Science and Technology Advisory Committee, chaired by the STAC Chairman and assisted by the Scientific Secretary. The General Assembly approves any EFC policy prepared and presented by the BoA. The Federation is managed through its General Secretariat with three shared headquarters located in London, Paris and Frankfurt.

The EFC carries out its most important activities through its nineteen active working parties devoted to various aspects of corrosion and its prevention, covering a large range of topics including: Corrosion and Scale Inhibition, Corrosion by Hot Gases and Combustion Products, Nuclear Corrosion, Environment Sensitive Fracture, Surface Science and Mechanisms of Corrosion and Protection, Physicochemical Methods of Corrosion Testing, Corrosion Education, Marine Corrosion, Microbial Corrosion, Corrosion of Steel in Concrete, Corrosion in Oil and Gas Production, Coatings, Corrosion in the Refinery Industry, Cathodic Protection, Automotive Corrosion, Tribo-Corrosion, Corrosion of Polymer Materials, Corrosion and Corrosion Protection of Drinking Water Systems, Corrosion of Archaeological and Historical Artefacts. The EFC is always open to formulating new working parties in response to the demands brought about by developing technologies and their ensuing corrosion requirements and applications.

The European Federation of Corrosion's flagship event is EUROCORR, the most important Corrosion Congresses in Europe, which is held annually in a different European country in September of each year. To date, 27 EUROCORR conferences have taken place in 12 different countries and they have gained a reputation for their high technical quality, global perspective and enjoyable social programme. Another channel for the EFC's valuable transfer of knowledge is the EFC "green" book series which are the fruit of the collaboration and high scientific calibre within and amongst the EFC working party members and are emblematic of the EFC editorial policy.

EFC Offices are located at:
European Federation of Corrosion, The Institute of Materials, Minerals and Mining,
1 Carlton House Terrace, London SW1Y 5DB, UK

Fédération Européenne de la Corrosion, Fédération Française pour les sciences de la Chimie, 28 rue Saint-Dominique, F-75007 Paris, France

Europäische Föderation Korrosion, DECHEMA e.V., Theodor-Heuss-Allee 25, D-60486 Frankfurt-am-Main, Germany

This recommended practice (RP) is a compendium of current best practices and state-of-the-art knowledge by major operators, engineering contractors and service companies involved in hydrocarbon production and transportation. It is intended to contribute to continued and trouble-free operation of pipelines. This enables the improved integrity of operations, safety and security of people whilst minimising the adverse impact on the environment and enabling security and business success.

The RP incorporates some minimum operational requirements and practices to ensure that when managing corrosion in pipelines, fundamental principles are followed. It covers management of corrosion for pipelines carrying hydrocarbons, injection water and/or produced water from design to decommissioning. It is structured to follow the logical steps of a basic corrosion management process and makes references to relevant and available international standards and/or recommended practices. This RP is intended for use by personnel from the petroleum industry having knowledge of corrosion and materials.

In the context of this RP, pipeline is defined as a line which carries fluids in upstream hydrocarbon production operations from wellheads or production sites to process sites (including platforms, reception plants, and downstream facilities) and vice versa. It excludes wells, vessels, rotating equipment and process pipeworks.

The recommendations set out in this RP are meant to provide flexibility and must be used in conjunction with competent technical judgements. Nevertheless, it remains the responsibility of the user of the RP to judge its suitability for a particular application or context.

If there is any inconsistency or conflict between any of the recommended practices contained in the RP and the applicable legislative requirement, the legislative requirement shall prevail.

Every effort has been made to ensure the accuracy and reliability of the data and recommendations contained in this RP. However EFC, its subcommittees, and individual contributors make no representation, warranty, or guarantee in connection with the publication or the contents of this RP and hereby disclaim liability of responsibility for loss or damage resulting from the use of this RP, or for any violation of any legislative requirements.

Pipeline Corrosion Management Working Group (WG) of the EFC Working Party (WP) 13 on Corrosion in Oil and Gas Production held its first meeting in June 2009 in Paris. Since then, several meetings were held to capture industry-wide recommended practices in corrosion management of pipelines transporting fluids, in hydrocarbon production. The organisation of the WG was undertaken by representatives from worldwide operators, engineering contractors and service companies.

In achieving the primary objective parameters affecting corrosion of pipelines, their mitigation and management were discussed during the WG meetings. These aspects formed the core of the present RP, sections of which have been prepared by the WG members. The RP was put together by Bijan Kermani of KeyTech through a joint industry project (JIP) to manage, collate inputs and prepare the document. The JIP was sponsored by BG Group, BP, ConocoPhillips, Shell, Statoil, Technip, Total, Wellstream and Wood Group Integrity Management (WGIM).

The chairmen of the WG and WP would like to express their appreciation to all who have contributed their time and effort to ensure the successful completion of this RP. In particular, we wish to acknowledge significant input from these individuals and their respective companies:

BG Group	Richard Carroll
BP	Alan Crossland
	Ammer Jadoon
	Lee Smith
ConocoPhillips	Mike Swidzinski
Shell	Sergio Kapusta
Statoil	Trygve Ringstad
	Helen Sirnes
	Bernt Slogvik
Technip	Carol Taravel Condat
	Elias Remita
Wellstream	Richard Clements
Wood Group Integrity Management (WGIM)	George Winning
	Chris White

Many members of EFC WP 13 also contributed to the development, review and specific sections of the RP whose significant contributions are recognised. In particular, efforts in reviewing the document from Mohsen Achour (ConocoPhillips), Mehdi Askari (Pars Oil and Gas Co), Ilson Palmieri (Petrobras), Ulf Kivisakk (Sandvik), Sytze Huizinga (Shell), and Michel Bonis (Total) are highly appreciated.

Finally, one editor (BK) wishes to thank JIP members for their financial support allowing development and production of this publication.

It is hoped that this RP will prove to be a key reference document for engineers, suppliers and contractors working in the oil and gas industry, paving the way for corrosion-free operation of pipelines with the ultimate goal of improving safety, security and minimising the impact on the environment.

Bijan Kermani	Thierry Chevrot
KeyTech	Total
Chairman of Working Group	Chairman of EFC Working Party
Pipeline Corrosion Management	Corrosion in Oil and Gas Production

Pipeline corrosion management (PCM) is defined as that part of the overall pipeline management system (PMS) and operating practices which is concerned with the development, implementation, review and change of corrosion policy. In this recommended practice (RP), PCM covers the management of corrosion from design to decommissioning, for upstream pipeline systems carrying hydrocarbons, injection water and/or produced water. While numerous instances of pipeline leak and failure are reported annually, the challenge to managing their occurrences has not been reflected in the production of many international standards. This Chapter focuses on introducing the subject, outlines the objectives encompassing the scope of the RP and the benefits from the implementation of the PCM programme. It underlines necessary management commitment and competency requirements.

1.1 Foreword

Pipeline integrity is key to maintaining operational success, safety and security and minimising harm to the environment. Several industry surveys [1,2] have shown that corrosion is a dominant contributory factor to failures, leaks and integrity threat in pipelines. Therefore, its optimum control within an integrity management framework is paramount for the cost effective design of facilities and ensuring continued, uninterrupted and safe operations within the expected design life.

This RP is developed to improve and manage corrosion in upstream pipeline systems. It focuses on corrosion exclusively; both internal and external. However, other integrity threats should be considered and addressed within an appropriate Asset Integrity Management (AIM) programme to enable integrity and reliability assurance.

The risk of internal or external corrosion becomes real once an aqueous phase is present and able to contact the pipe wall, providing a ready electrolyte for corrosion reactions to occur. The inherent corrosivity of this aqueous phase is then heavily dependent on the pipeline materials of construction, environmental conditions (temperature, pressure, presence of bacteria, etc.) and levels of dissolved corrosive species (acid gases, oxygen, organic acids, etc.) which may be present.

Development, implementation and continual updating of a successful corrosion management framework are key components of operating a pipeline throughout its production life, whatever the operating conditions. As a consequence, it is important that an adapted and tailor-made corrosion management programme is put in place to actively manage corrosion from the earliest stages through to the end of pipeline life to ensure trouble-free operations. Such an approach needs to be taken on board and implemented in the technical/commercial assessment of new field developments, in prospect evaluation, and for operating existing pipeline systems.

The RP provides overview of elements, steps and measures necessary to manage corrosion of pipelines in oil and gas production and transportation. It has been developed from feedback of operating experience, research outcomes and operating companies' in-house studies and practices. It has captured extensive inputs from operating companies, manufacturers, engineering contractors and service providers.

This RP is intended for use by personnel from the petroleum industry having knowledge of corrosion and materials and the necessary competency described throughout this document.

In the context of this RP, pipeline is defined as a line which carries fluids from wellheads or production sites to process sites (including platforms, reception plants, and downstream facilities) and vice versa, excluding wells, vessels, rotating equipment and process pipework.

This RP is not intended to provide a prescriptive framework for PCM, but only to outline, without going into details, the techniques, steps or measures which have been demonstrated as successful in the identification and management of the corrosion threats to oilfield pipeline facilities. It will also make reference to existing and recognised international standards whenever possible, without further details of their contents.

Furthermore, the present RP only provides a partial answer focused on corrosion threats of what needs to be developed within a broader theme of AIM.

By following the procedures outlined in this RP, confidence can be gained that the levels of corrosion control and management are appropriate for safe and trouble-free operations.

1.2   Objectives of this RP

This RP aimed to produce a reference document to achieve the following objectives:

- To capture and compile into a comprehensive document industry-wide information in relation to PCM.
- To define and outline common practices in PCM available amongst oil and gas operating companies and engineering design houses to:
  - ease discussions with projects' partners worldwide
  - deal with third party tie-ins.
- To agree with engineering design houses on processes and approaches necessary to define and characterise corrosion management requirements at the design stage.

The RP should be treated as an integral part of a pipeline management system (PMS).

This RP includes:

i.   Recommendations for registering essential and/or useful pipeline data for implementing adequate corrosion management programmes.
ii. An overview of corrosion threats and their respective assessment practices.
iii. Recommendations for defining and implementing corrosion mitigation actions.
iv. An overview of corrosion monitoring data capture and full analysis.
v.   An explanation for the need of adequate inspection programmes and their contribution to PCM programme.
vi. Guidelines for an overall review process based on points (i) to (v) above.
vii. Guidelines for change of duty if necessary from the review in step (vi).

The recommendations set out in this RP are meant to provide flexibility and must be used in conjunction with competent technical judgement. It remains the responsibility of the user of the RP to judge its suitability for a particular application.

Throughout this RP, the terms 'should' is used to indicate recommendations as opposed to prescriptive requirements.

1.3   Scope

1.3.1   General

This RP covers requirements and gives recommendations for managing corrosion from design to decommissioning in pipeline systems carrying hydrocarbons, injection water and/or produced water. It incorporates some minimum operational requirements and practices to ensure that fundamental principles are followed when managing corrosion in pipelines. It is intended for use in the technical/commercial assessment of new field developments, in prospect evaluation, design and construction, and for operating existing pipeline systems.

Table 1.1 and Fig. 1.1 summarise the scope of this RP, and present a list of excluded items/equipment where applicable. It should be noted that items within this RP include pig launchers and receivers.

Table 1.1   Scope of this RP

Items covered by this RP[1,2]	Excluded items
Risers, transportation pipelines and trunklines for liquids, gases and multiphase fluids	• Lines handling gas prepared for domestic use (gas distribution systems) • Lines handling processed products • Valves and rotating equipment • Subsea templates • Pressure vessels • Topside/surface facilities • Compression stations • Utility lines/umbilicals/chemical injection lines • Drain lines • Non-metallic materials • Slug catchers[3]
Flowlines and gathering lines	Crude oil storage, handling facilities, vessels and rotating equipment, heat exchangers, field facilities and processing plants
Manifolds and pipeline end manifolds (PLEMs)	Wellhead template
Catenary/dynamic risers	J tube, Caissons
Pig launchers and receivers	Valves
Flexible pipelines	
Corrosion-resistant alloy (CRA) and CRA clad/lined and non-metallic lined pipelines	
Injection lines (CO_2 injection, steam lines)	Pumps, chemical injection lines, steam generation, CO_2 removal plant

[1] In the context of this RP, flowlines, gathering lines and trunklines are considered as pipelines.

[2] Items within this RP include pig launchers and receivers.

[3] While slug catchers are designed within the pipeline codes, they are project specific and therefore not included in this RP.

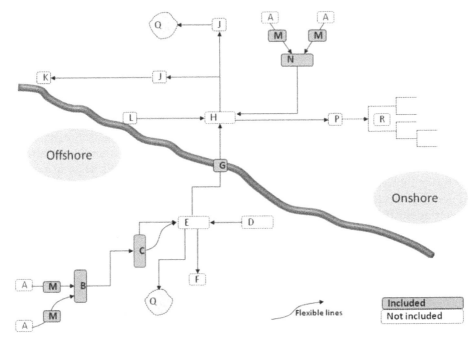

1.1 A schematic diagram of typical pipeline systems covered by this RP.
Solid lines: Items included within the scope of this RP.
Dashed lines: Items excluded from the scope of this RP.
Items within the scope of this RP include pig launchers and receivers.
A, Individual producing well. B, Subsea manifold. C, PLEM. D, Subsea template.
E, Offshore producing/processing facilities. F, WI/WAG wells. G, Landfall. H,
Onshore processing facilities. J, Compressor station. K, Export terminal. L,
Import terminal. M, In-field flowlines. N, In-field manifolds.
P, Let-down station. Q, Gas storage field. R, Gas distribution network

1.3.2 Contents

The approach adopted in this RP follows the systematic steps taken by most operating companies in the implementation of PCM. It follows the logical steps of a basic corrosion management process and is structured accordingly. This includes assessment, mitigating measures, design and implementation, monitoring and inspection, review and change of duty. It makes reference to relevant international standards and/or recommended practices when available.

The structure of the document is captured in Fig. 1.2 with a brief description of each section summarised in Table 1.2. At the end of each Chapter or Annex, a section is assigned to relevant bibliography where additional detailed and pertinent information can be found. A more detailed description of each topic is given at the start of each Chapter or Annex.

1.4   Purpose of Pipeline Corrosion Management (PCM)

A 'live' PCM programme will act as an effective way to control the corrosion threats in pipelines. The purposes of developing, implementing and continuously updating the PCM programme are to:

- Prevent loss of containment caused by corrosion.
- Provide a management directive with the purpose of controlling corrosion threats and costs over the lifetime of the pipeline systems.
- Ensure that adequate corrosion control programmes are in place and reviewed systematically.
- Define clear responsibility, accountability and ownership of corrosion management throughout the operating life of pipelines.

In addition, a PCM programme should ensure compliance with national and international laws and regulations.

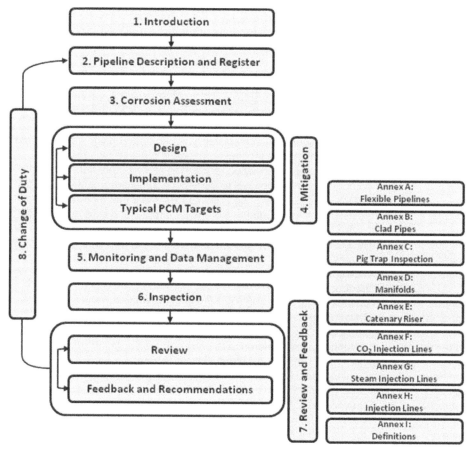

1.2 Structure of this RP

1.5   Benefits of PCM programme

Development and implementation of a structured corrosion management programme will ensure continued and safe operations of pipelines. In addition, key benefits arising from the development and implementation of PCM programme include but are not limited to:

- balancing capital expenditure (CAPEX) and operating expenditure (OPEX)
- optimising materials selection and corrosion control measures

- minimising any adverse effect on the environment
- providing safety and security
- understanding of corrosion activities in pipeline systems by:
  - providing an early warning of active corrosion which enables cost-effective and early intervention
  - directing towards mitigation and monitoring/inspection measures
  - confirming effectiveness of remedial measures
- enabling better-informed integrity, maintenance and repair planning decisions
- providing feedback on operational conditions leading to corrosion and tailoring corrosion modelling activities.

Overall, PCM together with risk-based assessment (RBA) methodologies within the context of AIM will lead to minimising loss of containment, as well as optimising future repairs and inspection needs. This can result in potential cost savings.

Table 1.2   Contents of this RP

Chapter/section	Description
Chapter 1	Introduction outlining the scope, purpose of PCM, objectives, benefits and applicability of this RP
Chapter 2	Pipeline description and register covering basic design data required for assessing and managing corrosion in pipeline systems
Chapter 3	A description of information required to assess corrosion, competency level required to carry out such assessments, types of threats and overview of corrosion assessment methodologies
Chapter 4	Key elements of mitigation actions for internal and external corrosion control
Chapter 5	Necessity and means for corrosion monitoring and data gathering
Chapter 6	Necessity for inspection and its contribution to PCM
Chapter 7	Highlighting steps, measures and manners to review and provide feedback
Chapter 8	Change of duty and implementation of change
Annex A	Covering specific reference to corrosion management of flexible pipelines
Annex B	Covering specific reference to corrosion management of clad pipes, liners and corrosion-resistant alloy (CRA) lines
Annex C	Covering specific guidance on pig trap inspection and maintenance
Annex D	Covering specific reference to corrosion management of manifolds
Annex E	Covering specific reference to corrosion management of catenary risers
Annex F	Covering specific reference to corrosion management of CO_2 injection lines
Annex G	Covering specific reference to corrosion management of steam injection lines
Annex H	Covering specific reference to corrosion management of injection lines
Annex I	Definitions

1.6   Management commitment

The implementation, use, and maintenance of a suitable PCM programme require adequate resources and therefore, management commitment.

Typical ways of demonstrating management commitment may include, but are not limited to:

- The preparation of multi-annual budgets dedicated to the PCM to enable middle/long term planning.
- Regular (annual) management reviews of the effectiveness of the PCM as part of the overall AIM programme.
- The formal approval of a corrosion management policy.
- The maintenance of competencies and skill levels in the team responsible for PCM, which can be encouraged by:
  - rewarding career path
  - establishing adequate continuity through succession plans
  - recognising the expertise of key personnel
  - ensuring the implementation of specific competency development programmes.

1.6.1   Responsibilities/accountabilities

Corrosion management in general is an activity requiring the involvement of several disciplines. In the case of PCM, it is necessary that responsibilities and accountabilities of the different entities and personnel involved are clearly defined and documented to ensure that:

- all activities required by the PCM are adequately addressed and carried out
- the interfaces between the different entities are clear
- the PCM programme is followed rigorously and efficiently, thereby avoiding duplicated tasks
- each task related to the PCM programme is carried out by personnel with adequate qualifications and experience.

The easiest and most reliable way of documenting responsibilities and account-abilities is to produce an interface matrix such as the template presented in Table 1.3. This is typically represented in a 'Responsible Accountable Consulted Informed' (RACI) format. This RACI describes the participation of various personnel in completing tasks or deliverables for a particular pipeline system. This is particularly useful in clarifying roles and responsibilities in cross-functional/departmental projects as well as processes involved in PCM.

1.6.2   Corrosion management policy

A corrosion management policy is essential to clearly document:

- management and personnel commitment towards the PCM
- high level objectives of the PCM
- performance targets
- means and resources to achieve these objectives.

The corrosion management policy can be complemented by other documents giving more detailed information and directives about the various tasks to be achieved and procedures to be followed. An example of such documents are outlined in the Bibliography Section (Section 1.10). These may be related to a variety of topics which may, for example, include: corrosion assessments, corrosion monitoring, chemical treatment, maintenance of protective coatings, etc.

Table 1.3 A typical RACI template

	The Leadership Team	Engineering Authority	Technical Authority	Field Operation	Others
Business planning					
Corrosion strategy & planning					
Corrosion management programme implementation					
Assessment and reporting of corrosion management programme results					
Conformance with corrosion management programme					
Exception to corrosion management strategy					
Exception to corrosion management programme					
Others					

1.7   Optimisation and use of corrosion modelling

The philosophy behind this RP is to reinforce the importance of methodologies, corrosion mitigation actions, acquiring feedback and using operating company's experience to achieve effective PCM. It is therefore recognised that, while useful for corrosion assessment, corrosion models in isolation cannot be relied upon solely to implement a satisfactory PCM programme. In this context, corrosion models should be used bearing in mind corrosion monitoring/inspection data, historical events, operational trends, past experience and field data.

Furthermore, corrosion models should be used by personnel having an overall understanding of corrosion management, to take into account their limits and their relevance with regard to field conditions and experience.

1.8   Competency of personnel

The competency of personnel involved in corrosion management activities is of paramount importance to the success of PCM. While this will normally be handled through the corporate policy of the operating company, the following aspects should be underlined:

- It is recommended that key positions are identified in the PCM programme to ensure that they are staffed on a permanent basis, and that a multi-annual recruitment and training plan is made available.
- Personnel should have received adequate training to undertake the tasks they will be assigned to in the PCM programme. Training should be adapted to these tasks and should involve courses and/or on the job training.

- It is of importance that all personnel involved in PCM, in whatever capacity, spend sufficient time in the field. This should be either through an assignment or site visits to become familiar with operational practices. It is reinforced that PCM should not solely be a 'paper exercise', as many deficiencies can only be identified on site through operational involvement.
- It is also important that personnel involved in PCM are suitably motivated. This will generally be covered by corporate policies. Achievements in preserving pipe-line integrity contribute to the operating company's business objectives and public image. This should be recognised adequately if the operating company wants to maintain sufficient expertise and experience in their PCM personnel.

1.9 Abbreviations

AC	Alternating Current
AIM	Asset Integrity Management
ALARP	As Low As Reasonably Practicable
BOL	Bottom of Line
C-steel	Carbon and Low Alloy Steel
CA	Corrosion Allowance
CAPEX	Capital Expenditure
CCS	CO_2 Capture and Storage
CI	Corrosion Inhibitor
CIS	Close Interval Survey
CP	Cathodic Protection
CRA	Corrosion-Resistant Alloy
CUI	Corrosion Under Insulation
DC	Direct Current
DCVG	DC Voltage Gradient
ECDA	External Corrosion Direct Assessment
EMAT	Electromagnetic Acoustic Transducer
ER	Electrical Resistance
FAT	Factory Acceptance Test
FPCM	Flexible Pipeline Corrosion Management
GAB	General Aerobic Bacteria
GLR	Gas Liquid Ratio
GnAB	Gram-negative anaerobic bacteria (*GNAB*)
GOR	Gas Oil Ratio
GPS	Global Positioning Systems
HDD	Horizontal Directional Drilling
HDPE	High-Density Polyethylene
HE	Hydrogen Embrittlement
HIC	Hydrogen Induced Cracking
HV	High Voltage
ILI	Inline Inspection
JIP	Joint Industry Project
KPI	Key Performance Indictor
LPR	Linear Polarisation Resistance
MAOP	Maximum Operating Pressure
MAWT	Minimum Allowable Wall Thickness

MDEA	Methyl-Di-Ethanol-Amine
MFL	Magnetic Flux Leakage
MIC	Microbial Induced Corrosion
NDE	Non Destructive Evaluation
OPEX	Operating Expenditure
Ox Sca	Oxygen Scavenger
P	Pressure
PA	Polyamide
PCM	Pipeline Corrosion Management
PE	Polyethylene
PEX	Cross-Linked Polyethylene
PIM	Pipeline Integrity Management
PIMS	Pipeline Integrity Management System
PIP	Pipe in Pipe
PLEM	Pipeline End Manifold
PLET	Pipeline End Termination
PMS	Pipeline Management System
ppb	Parts per billion
ppm	Parts per million
PWHT	Post Weld Heat Treatment
RACI	Responsible Accountable Consulted Informed
RBA	Risk Based Assessment
RBI	Risk Based Inspection
RIM	Riser Integrity Management
RP	Recommended Practice
SCC	Stress Corrosion Cracking
SCR	Steel Catenary Riser
SOHIC	Stress Oriented Hydrogen Induced Cracking
SRB	Sulphide Reducing Bacteria
SSC	Sulphide Stress Cracking
SWC	Stepwise Cracking
T	Temperature
TDS	Total Dissolved Solid
TOLC	Top of Line Corrosion
TR	Transformer Rectifier
UT	Ultrasonic
VOC	Volatile Organic Compounds
WG	Working Group
WI	Water Injection
WP	Working Party

Bibliography

- UK Health and Safety Executive, 'A guide to the pipelines safety regulations, guidance on regulations', 1996.
- Energy Institute, 'Guidance for corrosion management in oil and gas production and processing', May 2008.
- Canadian Association of Petroleum Producers (CAPP), 'Mitigation of internal corrosion in sweet gas gathering systems', June 2009.

- Canadian Association of Petroleum Producers (CAPP), 'Mitigation of internal corrosion in sour gas pipeline systems', June 2009.
- Canadian Association of Petroleum Producers (CAPP), 'Mitigation of internal corrosion in oil effluent pipeline systems', June 2009.
- Canadian Association of Petroleum Producers (CAPP), 'Mitigation of internal corrosion in oilfield water pipeline systems', June 2009.
- UK Health and Safety Executive, 'A guide to the pipelines safety regulations', 1996.
- 'Pipeline risk management manual – ideas, techniques and resources', 3rd edn, (ed. W. K. Muhlbauer), 2004, Elsevier.

References

1. P. Zelenak, H. Haines and J. Kiefner: 'Analysis of DOT reportable incidents for gas transmission and gathering system pipelines, 1985 through 2000', PR218-0137, Kiefner and Associates Inc. for the Pipeline Research Council International (PRCI), 2003.
2. http://events.nace.org/library/corrosion/pipeline/pipeline-failures

An essential first step in managing corrosion in any pipeline system is to compile an exhaustive inventory of relevant information into a pipeline register. Such a list characterises the pipeline and should include (i) all details related to the pipeline location, field and history, (ii) design details including pipe configuration, materials data, (iii) operating parameters covering types of fluid and environment (burial, atmospheric, marine, immersed, soil if known, etc.), operating conditions, acid gas content and water chemistry, (iv) internal and external corrosion mitigating measures covering operational pigging, chemical treatment, type of cathodic protection (CP) system, coatings, and (v) inspection strategy including pigging capacity and other related information. This chapter (see Fig. 2.1) outlines necessary information required to capture the full inventory of pipeline systems, their compilation into the pipeline register and provides a snapshot of steps and elements in this process.

2.1 Introduction

An overriding parameter in pipeline corrosion management (PCM) is full knowledge of pipelines for which an operating company is responsible. In some instances, this inventory and register are required by local legislations. This knowledge is gained through development and updating of a full pipeline register containing all related physical, logistical and operational parameters and information. Such information will also contain the basic data necessary to carry out and implement the respective corrosion management programme.

2.2 Basic pipeline register (the system)

The basic pipeline register aims to document a full list of pipelines for which an operating company is responsible for their corrosion management. It should be noted that this may also include pipelines not necessarily owned by the operating company, for example, in the following cases:

- The pipeline or part thereof, is owned by a third party which delegates its operation to an operating company/contractor.
- The pipeline or part thereof, is owned by a consortium comprising the operating company and has clearly delegated its corrosion management to the said operating company.
- While the pipeline or part thereof is operated by a third party, its corrosion management has been delegated – such a case may apply to third party tie-ins which may include complementary requirements and additional information.

The pipeline register should contain basic information required to document the characteristics of the lines ('hardware'), as well as the responsibility of the operating company with regard to corrosion management with a specific attention to interfaces when relevant.

The information listed in the basic pipeline register should include, as a minimum but may not be necessarily limited to those outlined in Table 2.1.

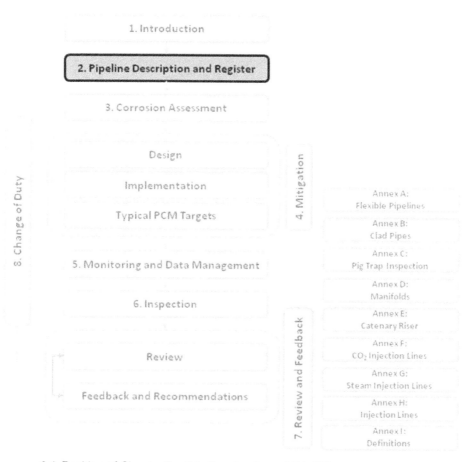

2.1 Position of Chapter 2 within the structure of this RP

2.3   Physical characteristics register

Basic design and construction data need to be documented in the pipeline register. This is necessary to assist in any remaining life evaluation by corrosion assessments. The accuracy of such information is paramount, since the remaining life of a pipeline is substantially affected by critical aspect such as materials, nominal wall thickness, type of coating, or the method used for CP.

The basic design and construction data which should be documented comprises, but may not be limited to those outlined in Table 2.1.

In some specific cases, additional information may be documented such as the names of the linepipes' supplier(s), welding and/or field joint coating application contractor(s), design and installation contractor(s), etc. This type of information may be particularly useful in the case of very long pipelines where several contractors may have been involved.

Table 2.1 Typical content of pipeline register/inventory

Type of data	Information	Essential (Yes/No)
Pipeline system	Name of the owner	Yes
	Name of the operating company	Yes
	Battery limits	Yes
	Entity responsible for corrosion management	Yes
	Country, field, sector, asset	Yes
	Type (rigid, flexible, pipe in pipe (PIP))	Yes
	Departure and arrival points	Yes
	TAG Number	No
	Offshore, onshore, or both	Yes
	Diameter and length	Yes
	Year of commissioning	Yes
	Share of production of asset/operating company	No
	Design life	No
	Land ownership	Yes
Physical characteristics	Material & grade	Yes
	Wall thickness	Yes
	Design pressure	Yes
	Maximum operating pressure (MAOP)	Yes
	Minimum allowable wall thickness at MAOP	Yes
	Type and thickness of external coating	Yes
	Type and thickness of internal coating	Yes
	Type and thickness of field joint coating	Yes
	Internal or external cladding & thicknesses	Yes
	External cementing	Yes
	Type of cathodic protection system	Yes
	Operational pigging capabilities	Yes
	Topography	No
	Inline inspection (ILI) capabilities	Yes
	Permanent/temporary pig traps	No
	Laying (soil, seabed, trenched, buried, etc.)	Yes
	Landfalls	Yes
	Linepipe supplier(s)	No
	Welding & field joint coating contractor(s)	No
	Design & installation contactor(s)	No
Operating conditions	Physical operating conditions	
	• *Inlet & outlet pressures*	Yes
	• *Inlet & outlet temperatures*	Yes
	• *Oil, gas, condensate, water flowrates*	Yes
	• *Ambient temperature*	Yes
	Fluid characteristics	
	• *Type of fluid*	Yes
	• *Gas oil ratio*	Yes
	• *Inlet & outlet water flowrates*	Yes
	• *CO_2, H_2S contents (gas phase)*˙	Yes
	• *Dissolved O_2 content*	Yes
	• *Typical water composition*	Yes
	• *Organic acid content*	Yes
	• *Solid content*	Yes

Table 2.1   Continued

Type of data	Information	Essential (Yes/No)
Inspection data	Inspection policy (prescriptive, risk based inspection (RBI), etc.)	Yes
		Yes
	Date and type of previous inspection	Yes
	Date and type of next inspection	No
	High level summary of last inspection results	Yes
	Repairs (composite repair, clamps)	Yes
	Pipeline section replacement	
Others	Physical treatments & level of confidence	
	• *De-aeration*	Yes
	• *De-carbonation*	Yes
	• *De-sulphuration*	Yes
	• *De-hydration*	Yes
	• *Operational pigging*	Yes
	Chemical treatments & level of confidence	Yes
	• *Corrosion inhibition*	Yes
	• *pH stabilisation*	Yes
	• *Biocide treatments*	Yes
	• *Oxygen scavenging*	Yes
	• *Anti scale treatment*	Yes
	• *H₂S scavenging*	Yes
	• *Demulsifiers*	No
	• *Wax/asphaltene inhibitors*	No
	• *Elemental sulphur solvent*	Yes
	• *Mode of treatment*	Yes
	• *Dosage rate*	Yes
	• *Name of supplier*	Yes
		No
	Corrosion monitoring	
	• *CP data, close interval survey (CIS), DC voltage gradients (DCVGs), etc*	No
	• *Monitoring type & technique*	No
	• *Locations*	No
	• *Monitoring frequency*	No

* CO_2 and H_2S contents are the respective values of the last or next gas phase in equilibrium with the aqueous phase, i.e. the content at the last separator for any liquid circuit downstream, or the content at the bubble point for any hydrated oil upstream.

2.4   Operating conditions

The pipeline register should document the basic operating conditions of each pipeline to provide a sound basis for performing corrosion assessment. Operational conditions will also enable corrosion engineers to identify potential corrosion threats, their significance, and their implications in terms of mitigation actions and monitoring.

Operating conditions should be updated on a regular basis to reflect changes in pipeline operations so that the corrosion assessment remains valid with time. Therefore, the pipeline register becomes a live document.

The operating conditions can be divided into two parts as follows:

2.4.1 Physical operating conditions

As most corrosion mechanisms are affected by or sensitive to the operating pressures and temperatures (both internal and environmental), these parameters need to be documented in the pipeline register. Furthermore, the type of external environment (burial, atmospheric, marine, immersed, etc.) needs to be documented to ensure that a complete external corrosion assessment can be performed. Respective parameters are outlined in Table 2.1.

Further information, such as soil properties, ambient temperature or seawater currents may be useful to document for specific applications.

2.4.2 Fluid characteristics

These are of paramount importance for carrying out corrosion assessments, defining mitigation actions and monitoring of corrosion. They should be captured and incorporated into the pipeline register including those outlined in Table 2.1.

Organic acids are important both in terms of system corrosivity and especially for determining the risk of top of line corrosion (TOLC). The presence of solids as sand, mineral scales, etc.) also needs registering.

More details on fluid characteristic are given in Chapter 3.

2.5   Inspection policy and data

PCM is an integral part of integrity management and, as such, it cannot be dissociated from the prevailing inspection policy. Furthermore, the availability of inspection data will impact the corrosion assessment, or even the definition of mitigation and monitoring actions to a large extent.

It is therefore, recommended to clearly document in the pipeline register, both the inspection policy followed for each pipeline, as well as the main results obtained from inspection activities.

When doing so, one should remember that inspection activities are different from corrosion monitoring activities, i.e. they aim at obtaining data on the physical status of the pipeline, rather than evaluating fluid corrosiveness or corrosion inhibition effectiveness as described further in Chapters, 4, 5 and 6.

The pipeline register should therefore document those information outlined in Table 2.1. The type and methods of previous and next inspections should include ILI, excavations, visual inspections of risers, etc.

A summary of the results of previous inspections, although useful, is not required in the pipelines register. However, these should be easily accessible within the operating company's company document management system so that they can be accessed if required during corrosion assessments and reviews, as well as during the definition of mitigation and monitoring actions. More details of inspection policy are given in Chapter 6.

2.6   Complementary information

The following information may be documented so that it is readily available for corrosion assessments and corrosion reviews. If the information is not documented in the pipeline register, then it will need to be readily available elsewhere and gathered or summarised before performing any corrosion assessment or corrosion review. These include the following three series of parameters:

2.6.1 Physical treatments

The existence of physical treatment upstream of the pipeline inlet such as de-aeration, de-carbonation, de-sulphuration, or de-hydration will have a positive impact on system corrosivity as long as these processes are reliably and correctly operated. The use of such treatments and the confidence in their availability and reliability can be documented in the pipeline register but will in any case need to be readily available. This information is outlined in Table 2.1 and further details of such treatments are given in Chapter 4.

2.6.2 Chemical treatment

Similarly, any chemical treatment used for mitigating corrosion can be included in the pipeline register or be readily available elsewhere. These may include but not limited to those outlined in Table 2.1.

Furthermore, when documenting chemical treatments, additional information reflected in Table 2.1 should include the following:

- Is the treatment continuous, by regular batch, or occasional?
- Was the chemical tested and selected in the laboratory?
- What is the availability of the treatment on site?
- What is the dosage rate?
- Is the treatment monitored in terms of injection follow-up, and/or corrosion monitoring, and/or dedicated laboratory analyses?
- What are the chemicals suppliers and product types?

Documentation of the above will allow determination of an overall level of confidence in the mitigating actions when performing corrosion assessments.

Some chemical treatments such as demulsifier, wax and asphaltene removers can increase fluid corrosivity by promoting water wetting.

More details on chemical treatments are given in Chapter 4.

2.6.3 Corrosion monitoring data

Corrosion monitoring data (coupons, probes, bioprobes, cathodic protection performance, coating degradation, sample collection and analyses) constitutes another essential series of information necessary for corrosion assessments and corrosion reviews. While a brief summary or anomaly reporting may be included in the pipeline register, exhaustive documentation of corrosion monitoring data may be too large to be included, but will need to be readily available in a separate dedicated company data management system.

However, it will be possible to document in the pipeline register whether corrosion monitoring is used. In this case, the necessary information should include those outlined in Table 2.1.

Further details on corrosion monitoring are given in Chapter 5.

2.7   Key messages

Table 2.1 gives the typical and brief content of the pipeline register recommended in this RP. The final register used by the operating company may include other information, depending on local regulations, operating company integrity assurance policy, and availability of data.

Information considered as essential is suggested in the table, but the responsibility to include in the register all of the required data is ultimately the responsibility of the operating company.

Pipelines experience many types of corrosion damage, internally and externally including general and localised corrosion. These damage mechanisms are influenced and governed by physical, chemical and hydrodynamic aspects of the pipeline system. Identification of key corrosion threats is a fundamental step in pipeline corrosion management (PCM) and is an iterative process throughout the design, operation and service life of pipelines. This chapter (see Fig. 3.1) covers types of corrosion threat, information required to carry out corrosivity assessments, hotspot identification and system corrosivity evaluation for both routine and non-routine operations. In addition, the chapter includes considerations on competency levels required for carrying out corrosivity assessment.

3.1   Introduction

Recognition of corrosion threats yields key information allowing effective and pertinent mitigating actions. In assessing corrosion, the type of damage, and its initiation and development throughout service life are important considerations in ensuring implementation of PCM. Information necessary to perform a corrosion assessment is wide ranging and the subject of many documents and standards as outlined in the Bibliography.

3.2   Relevant corrosion threats

The main corrosion threats in pipeline systems are outlined in Tables 3.1 and 3.2 with relevance to service and primary influencing parameters, for internal and external corrosion, respectively. The threats listed in these tables are broadly in line with time dependence threat categorisation described in ASME B31.8S. Further details can be found in the Bibliography.

3.3   Gathering data input

Before undertaking the corrosion assessment, a list of key data should be gathered. The most commonly used information for corrosion assessment includes those data outlined in Tables 3.3 and 3.4 for internal and external corrosion, respectively.

It is important to recognise that any sample taken is only representative of the process stream at the point where it has been taken. Changes in the fluid chemistry between the inlet and sampling point, or during the time between sampling and analysis should be understood and accounted for in interpreting the results.

During the operating life of a pipeline, the fluid chemistry and gas content may vary with process upsets and/or operational requirements. Therefore, the frequency of water and gas sampling programmes needs to be adequate and representative and their trends with time recognised.

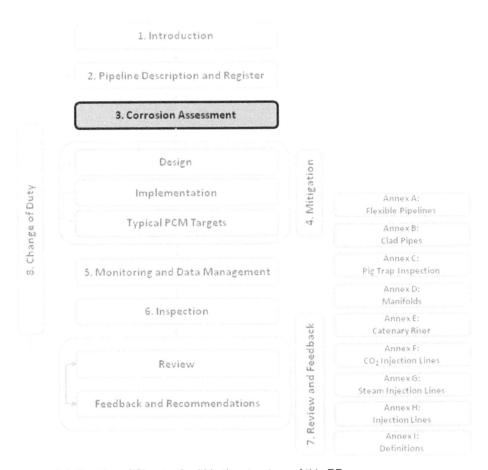

3.1 Position of Chapter 3 within the structure of this RP

Data relating to external condition tend to be collected less frequently, but are no less important and should still be subject to regular update and analysis. These may include but not limited to the soil and groundwater conditions, environmental conditions, and location (risers, landfalls, etc.).

3.3.1 Fluid chemistry

Representative water chemistry should be made available. This is critical for an effective internal corrosion threat assessment.

Key components in water chemistry for CO_2 and/or H_2S corrosion and for water injection lines are outlined in Table 3.3.

The following should be considered when carrying out fluid chemistry analysis:

- pH measurements made on fluids in the field on depressurised samples should be taken with caution. CO_2 evolves from the solution rapidly when the fluid is depressurised into a sample vessel. Therefore, the measured pH is not representative of the in-situ conditions.

Table 3.1   Internal corrosion threats

Type of threat	Relevance to service	Primary influencing parameters
Microbiologically influenced corrosion (MIC)	Oil and produced/ injection water	Sulphate, nutrients, bacteria type, pH, flow rate, salinity, temperature
CO_2 corrosion	Oil, gas and PW/PG injection	P&T, acid gas content, in-situ pH, flow regime, water cut
$H_2S + CO_2$ corrosion	Oil, gas and PW/PG injection	P&T, acid gas content, in-situ pH, flow regime, water cut, acid gas ratio, microbial activities
Top-of-line corrosion (TOLC)	Stratified wet gas	T, organic acid content, water condensation rate
Oxygen corrosion and other oxidants	Water injection	O_2 content, T and flow rate, residual chlorine
Under-deposit corrosion	Oil and produced/ injection water	Solids, scales, flow rate
Preferential weld corrosion	Oil, gas and PW/PG injection	Weldment metallurgy, flow regime, type inhibitor, water salinity
Crevice corrosion	Oil, gas and PW/PG injection	Joint design and configurations, oxygen content
Galvanic corrosion	Water injection	Dissimilar joints, oxygen content
Erosion and corrosion–erosion	Oil, gas and PW/PG injection	Solids, flow regime
Corrosion and cracking from injected chemicals[1]	At chemical injection point	Design of injection point, chemical dose rate, acid type, oxygen concentration[2]
Sulphide stress cracking (SSC)	Oil, gas and PW/PG injection	In-situ pH, H_2S, T, steel metallurgy, stress level
Hydrogen induced cracking (HIC)	Oil, gas and PW/PG injection	In-situ pH, H_2S, T, P, steel cleanliness and metallurgy
Corrosion-fatigue	Dynamic risers	Cyclic stress, system corrosivity

[1] Injected chemicals may contain contaminants that exacerbate the corrosion process, e.g. hydrate inhibitor may result in sufficient oxygen contamination to pose oxygen related damage in produced fluid lines.
[2] In highly aggressive environments, can also be a corrosion threat downstream of hydrate inhibitor injection points for wet gas/multi-phase systems, particularly where synergy with other corrosion threats is possible.

- Presence of some production chemicals may affect water chemistry. If anticipated, water samples should be taken upstream of injection point, before batch treatment or after injection has ceased for sufficient time.
- For samples taken during well tests, care should be taken to look for contamination with drilling fluids. High levels of ions such as Zn^{2+}, Ca^{2+}, Br^-, Ba^{2+}, Sr^{2+}, B^{3+} or K^+ may indicate contamination and therefore affect water chemistry.
- The organic acid content (or associated organic species) may interfere with bicarbonate analysis – a key point when carrying out water chemistry analysis. Care should be taken not to overestimate in-situ pH and bicarbonate level in the presence of organic acids.

Table 3.2 External corrosion threats

Type of threat	Relevance to service	Primary influencing parameters
Microbiologically influenced corrosion (MIC)	Buried pipelines	Soil type, coating, cathodic protection (CP) inefficiency, anaerobic conditions
Crevice corrosion	Above ground pipe support, flanges	Joint and support design
Galvanic corrosion	Flanges, couplings, fittings, insulating joints	Dissimilar joints, water wetting
Soil corrosion	Land lines	Soil type, coating, CP inefficiency
Atmospheric corrosion	Above ground lines	Coating, environment
AC-induced corrosion	Buried land lines	Third party infrastructures, telluric activities
Corrosion under insulation (CUI)	Above ground insulated lines	Type of insulation, the environment
Stray current corrosion	Buried land lines	Third party infrastructures
Stress corrosion cracking (SCC) – high pH	Buried land lines	Soil conditions, T, coating degradation, CP, pH, cyclic stress
Stress corrosion cracking (SCC) – neutral pH	Buried land lines	Coating degradation, CP, pH, cyclic stress
Corrosion-fatigue	Submarine lines	Cyclic stress
Hydrogen embrittlement	Submarine lines	CP overprotection, coating, steel/weldment hardness

- Sulphides may be measured in the aqueous fluid samples taken. However, H_2S can evolve from the sampling vessel on depressurisation and oxidise when exposed to oxygen. Therefore effective preservation of samples must be ensured. This is usually achieved by using a stabilising agent such as sodium hydroxide or zinc acetate [2].
- Operational procedures with regard to open and closed drains may lead to bacterial activities, water quantity and compositional changes – these should be taken on board.
- In the presence of Fe^{2+}, a suitable sampling and preservation (acidification or chelating agents) procedure should be used to prevent iron precipitation.
- It is recommended to drain the sample point for a sufficient period of time before collecting the fluids to ensure representative sampling.

3.3.2 Gas composition

The gas analysis should include the percentage of the acid gases (CO_2 and H_2S) and O_2. In the case of liquid pipelines, these are the values measured or predicted from the associated gas phase of the liquid/gas equilibrium immediately upstream.

3.3.3 Flow regime

Understanding the flow regime within the pipeline can help to assess system corrosivity, partitioning behaviour of an inhibitor, the likelihood of water wetting, under

Table 3.3 Summary of information for assessing internal system corrosivity

Parameter	Essential	Comment
Type of service	√	Oil, water, gas or multi-phase including any process changes
CO_2 content	√	In the associated gas
H_2S content	√	In the associated gas
Elemental sulphur	√	Very high corrosion rates can be experienced
O_2 content	√	Dissolved oxygen
Operating pressures	√	Range, maximum values
Bubble/dew point	√	
Operating & ambient temperatures	√	Range, maximum values
Oil API gravity		
Throughput/day gas	√	Profile throughout service life if available
Throughput/day oil	√	Profile throughout service life if available
Throughput/day water	√	Profile throughout service life if available
GOR/GLR		Profile throughout service life if available
Sand production and content		An important consideration for erosion and under deposit corrosion and considered essential for high flow rate conditions
Line internal diameter	√	
Flow regime		From modelling or meters/service knowledge
Water type & composition:	√	Condensed/formation water
HCO_3^-	√	
Organic acids and other organic species	√	e.g. acetate, formate, butyrate, etc. needed for CO_2, TOLC and MIC assessment
Cl^-		This is essential for flexible risers and CRAs
SO_4^{2-}		This is essential for MIC assessment
Fe^{2+}		
Ca^{2+}		
TDS		
Salinity		
pH	See footnote	Predicted and measured*
Bacterial count		For oil, multi-phase or water lines only
Operating life	√	
Wall thickness	√	
Remaining operating life		
Chemical treatment injection reliability		Meeting key targets in operation or are targets realistic in design?
Production chemicals applied		Meeting key targets in operation or are targets realistic in design?
Continuous or batch treatment	√	Meeting key targets in operation or are targets realistic in design? In the event of batch, frequency should be noted
Chemical residual content		Meeting key targets in operation or are targets realistic in design? This includes biocide, oxygen scavenger and corrosion inhibitor

GOR, gas oil ratio; GLR, gas liquid ratio; TDS, total dissolved solids; TOLC, top of line corrosion.

* It is essential that the pH is correctly measured or predicted to allow accurate system corrosivity assessment.

deposit corrosion and MIC. This is normally achieved by flow dynamic modelling bearing in mind operational parameters.

3.3.4   External assessment

Influential parameters affecting external corrosion of pipelines are summarised in Table 3.4.

The data should be gathered through a site survey covering variations along the pipeline. An initial desktop topographical, geological and environmental survey may be used to identify regions with fluctuations in water table, soil type and industrial activity. From this review, specific locations should be selected for site surveys and collection of samples for analysis or in situ measurement.

Soil resistivity is a key parameter and on its own, can provide a good assessment of the potential corrosivity of an environment. Typical values and guidance on the aggressiveness of the soils are given in Table 3.5. Resistivity measurements are described further in the Bibliography.

Preservation of soil samples taken in the field for analysis is no less critical than for fluid samples. Soil samples should not be contaminated or permitted to dry out, samples for bacterial analysis should be kept anaerobic, or sulphide reducing bacteria (SRB) tests conducted in the field should be stored under appropriate temperature conditions.

Table 3.4 Summary of information for assessing external system corrosivity

Parameter	Essential	Comment
Environment	√	Above ground, buried, submarine, splash zone, trenched. Landfall, crossings [1]
Geographical location and water depth	√	For sealines only
Exposure temperatures	√	Above and below pipe – winter and summer
Soil type/backfill	√	Consolidated, sand, gravel, etc.
Soil water content	√	
Bacteria level		SRB content per gram; aerobic, anaerobic
Coating/insulation/heating type	√	
Coating condition		Essential if condition unsatisfactory
Soil/electrolyte resistivity	√	
Soil pH		
Alkalinity		
Cl⁻		
SO₄⁻		
AC source interference	√	Railway, HV power, etc.
DC source interference	√	DC railways, other CP systems, industrial DC users
Cathodic protection type	√	Impressed current, sacrificial anode
Anode condition	√	
Pipe support design		
Mechanical type defects		Buckling, free-span, etc.

AC, alternating current; DC, direct current; HV, high voltage; SRB, sulphide reducing bacteria.

Table 3.5   Summary of Typical Soil Resistivity – Aggressivity Data

Resistivity	Corrosiveness	Typical soil conditions
Below 5 ohm.m (16 Ω.ft)	Very aggressive	Clay
5–10 ohm.m (16–33 Ω.ft)	Aggressive	Chalky
10–30 ohm.m (33–98 Ω.ft)	Moderately corrosive	Sandy
30–100 ohm.m (98–328 Ω.ft)	Mildly corrosive	
Above 100 ohm.m (328 Ω.ft)	Unlikely to be corrosive	

3.3.5   Existing third party activities and infrastructure

The key issues relating to PCM arising from third party activities may include:

- impact of third party fluids, physical and chemical mitigations on internal corrosion by tie-ins
- stray current effects (for DC voltage systems)
- AC corrosion due to proximity to AC voltage sources
- changes to the external chemistry/environment (e.g. industrial pollution).

AC corrosion is induced by proximity and parallelism to high voltage overhead power lines. This can be caused by either a long-term imbalance in the transmission system or high voltages near earthing (grounding) systems, resulting from lightning strikes and faults. Interference between the pipeline and power supply phases causes induced voltages and currents. Induced AC voltages can be very dangerous and induced AC currents can cause very rapid corrosion. Mitigation is therefore important to reduce both a safety and corrosion risk.

Data gathering should include those factors defined in Section 3.3.

It should be noted that encroachment of third parties may result in stray current risks or potential pollution of ground conditions leading to increased soil corrosivity.

3.4   Competency

Satisfactory execution of a complete corrosion assessment is a multi-disciplinary process. This implies that all relevant disciplines should be identified, be involved in the assessment and have adequate competency. Each participant should have a required combination of qualifications, training, experience and/or support/ supervision to ensure a reliable outcome. Competency can be gained by formal training or relevant experience. Examples of bodies offering training and/or certification are included in the Bibliography.

It is the responsibility of the operating company to ensure that the training programme is fit for purpose and adequately accredited. In a case of formal training or experience, this needs to be auditable to meet the requirement.

3.5   Non-routine operations

3.5.1   Well interventions

The most common non-routine operations are typically well interventions which may lead to backflow of well treatment fluids into the pipeline. Treatment fluids can be back-produced for a considerable period after the intervention. These fluids can be very corrosive to pipeline materials, often containing high concentrations of acids such as hydrochloric, hydrofluoric, acetic, EDTA, etc. The organic film forming inhibitors used for protection from CO_2 and/or H_2S corrosion will not normally offer protection against the low pH environment resulting from these acids. An additional threat with EDTA is a degradation threat to non-metallic components such as those used in seals or umbilicals. Backflow from a well intervention may also include solids, such as corrosion products or sludges which may be deposited within the pipeline and reduce the effectiveness of inhibition or biocide programmes.

It is critical to understand the influence of these operations and their frequency on the corrosion assessment process. Aspects to consider include:

- operations on acid flow back
- oxygen contamination
- microbial activities
- salt content
- solids and sludges.

3.5.2   Start-ups

Start-ups can lead to sand events for production well flowlines. From a corrosion management perspective, it is important to understand the correlation between production start-up rate and potential associated sand production. Apart from corrosion/erosion threats, sand production may also accumulate at low points of the pipeline and having an adverse effect on chemical treatment under such conditions.

3.5.3   Shutdown

Shutdown of pipelines for prolonged periods can result in increased risks in some systems. For example, seawater injection lines may suffer enhanced MIC risks due to increased time between biocide treatments.

3.6   Other hotspots

The hotspots, defined as locations of potential high corrosion threats, which are covered in this section, should be taken into account on a systematic basis during corrosion assessments. They are all project-specific and their assessment and mitigation should be addressed on a case-by-case basis.

3.6.1   Riser pipeline sections/splash zones

The marine splash zone is the area that is alternately immersed due to the action of waves and tides. The splash zone area varies based on the geographical location and its definition differs amongst operating companies.

In this area, corrosivity can be very severe, especially for hot risers (corrosion rates of 3–10 mm year^{-1} have been reported) and CP systems will not be effective. Corrosion protection for this section is totally dependent upon the condition of the coating system. The following issues and corresponding uncertainties should therefore be taken into account and documented:

- type and age of coating
- temperature
- potential for external damage of coating
- availability and quality of fabrication and installation documentation
- frequency and quality of external visual inspections including under-clamps and supports
- availability of other inspection results.

3.6.2   Touch down

This hotspot is covered in Annexes A and E.

3.6.3   Landfall

A landfall is a critical part of a submarine pipeline system going onshore which requires a thorough evaluation. The importance of a robust landfall must be stressed both during design and installation.

Design of landfall is covered by ISO 13623 and DNV OS-F101 (Appendix F). Relevant local regulations should be identified and evaluated due to possible legislative aspects.

Depending on the choice of design and local topography, corrosion assessment should take into account the presence of concrete culvert, pre-installed casings, gravel protection, wave breakers, etc.

Guidance

The following solutions are options in landfall design and each has a particular aspect that should be considered within PCM.

- open trench, back-filled after pipeline installation
- horizontal directional drilling (HDD)
- microtunnel
- blasted tunnel in rock with sealtube
- blasted tunnel in rock with direct pull-in of pipeline
- pre-installed casing-pipe, for later direct pull-in of pipeline.

A landfall represents a critical part of a submarine pipeline due to the following:

- high external corrosivity (splash zone area)
- no/limited effect from CP systems located subsea and onshore
- limited access for inspection and monitoring of external conditions
- limited or no access for repair once the pipeline has been installed and set in operation.

3.6.4   River/fjord/road crossing

Many aspects of corrosion assessment necessary for the design of landfalls apply equally to crossings. In addition, the following aspects should be taken on board:

- use of impressed current, sacrificial anodes or both along the pipeline – CP systems should not be mixed unless insulated
- use of insulating joints isolating incompatible CP systems
- accessibility and reliability of CP monitoring at the crossing.

Some specific recommendations related to CP design for such crossings are covered by ISO standard ISO 15589 Parts 1 and 2 and are further discussed in Chapter 5.

3.6.5   Insulating joints

Corrosion at the insulating joints should be considered. Some of the risks include:

- external short circuit due to burial
- external short circuit due to water flooding
- short circuit due to external instrumentation or support structures and earthing systems
- internal short circuit due to high water cut or improper design
- internal short circuit due to high voltage induced by lightening or welding.

Design, location and installation of the insulating joint should be reviewed to assess the potential impact of the likely damage mechanism.

It should be noted that if the fluid transported by the pipeline includes produced or injected water there is a risk of internal corrosion on the more negative CP potential side of the insulating joint. This is due to the conductivity and continuity of the water rather than its corrosiveness. To avoid such internal corrosion, the pipeline on the cathodic side (more negative potential) of the insulating joint should be lined with an electrically insulating material. The length of the section to be coated internally should be determined on a case-by-case basis.

3.6.6   Locations with high cooling rates

Locations with high cooling rates require special attention in the corrosion assessment process. Such locations include buckles exposing previously buried sections of pipelines, river/fjord crossings and cold seawater currents and thermal insulation damage areas. These may undergo top of line corrosion (TOLC) as referred to earlier in this chapter.

3.6.7   Risers in J-tubes

Corrosion assessment should consider the design of the J-tube and the resulting annulus management (open or closed annulus, treated or untreated annulus fluids, etc.).

3.7   System corrosivity

System corrosivity is governed by many parameters as outlined in Section 3.2 and also the nature of pipeline use, which generally falls into two broad categories of production or injection. For production, system corrosivity is associated with the presence of water and acid gases (CO_2 and H_2S) which are invariably produced with the hydrocarbon phase. In contrast, for injection, system corrosivity is governed by the nature of fluid being transported and primarily by bacteria and dissolved oxygen content – a specific case of PCM for injection lines is covered in Annex H.

This section describes the methods and procedures used in evaluating fluid corrosivity in pipeline systems, key elements of which are outlined in Table 3.6.

Many predictive models or criteria have been developed which attempt to correlate corrosion rate with service conditions. However, using these predictive models in isolation, there is a danger of working with a too simplistic or clear-cut picture of the situation. Experience is also required in the interpretation and implementation of the outcome. The following sections cover details of threats and their respective corrosion assessments.

3.7.1 CO_2 corrosion evaluation

Over the years, much effort has been expended in studying factors controlling CO_2 corrosion and the knowledge gained has led to the development of several prediction models. The main factors affecting CO_2 corrosion (namely partial pressure of CO_2, operating temperature and pH) are common to all these models, although the overlying effects of factors such as flow regime, film formation/deposition, hydrocarbon phase, realistic and representative testing, steel chemistry and corrosion inhibitor and above all field experience complicate the picture [3].

While some models have been developed to reflect such parameters, there is no international standard available as indicated in Table 3.6. These models have limitations and advantages [4]. Their widespread application is subject to ease of use, availability and applicability, although they are operating companies' preferences which have incorporated in-house experience and additional parameters.

3.7.2 H_2S + CO_2 corrosion evaluation

Corrosion assessment in H_2S containing media causing metal loss and cracking phenomena (Table 3.6) are as follows:

H_2S + CO_2 metal loss corrosion

As mentioned earlier, H_2S introduces additional complexity into CO_2 system corrosivity through the formation of FeS. There are no conclusive corrosion prediction models or assessment tools to predict the onset of FeS breakdown and localised attack. Most assessments are therefore, on operational experience and knowledge of the underlying CO_2 corrosion rate and CO_2/H_2S ratio [5,6].

H_2S assisted cracking

NACE MR0175/ISO 15156 international standard provides requirements and recommendations for materials qualification and selection for safe application in environments containing wet H_2S in oil and gas production systems. Some additional guidelines are provided in EFC Publications 16 and 17.

Table 3.6   Notable examples of corrosion evaluation methods

	Media	Corrosion threat	Manifestation	Principal parameters	Model and standard
Production	Hydrocarbon and water	CO_2 Corrosion	Metal loss (general and/or localised corrosion)	• CO_2 • Organic acids • Operating conditions • Water chemistry	Many models, operating company preference
	Hydrocarbon and water	$H_2S + CO_2$ Corrosion	Metal loss (general and/ or localised corrosion)	• CO_2/H_2S ratio • Operating conditions • Water chemistry	Subjective – no proven model
	Hydrocarbon and water	SSC/HIC	Different types of cracking	• H_2S • In-situ pH • Operating conditions • Water chemistry	NACE MR0175/ISO 15156
	Wet gas	TOLC	High pitting damage at 10–2 o'clock position	• Stratified flow regime • Cold spots • Operating temperature • Water condensation rate • Organic acid content	Rules of thumb, limited number of models available
	Multiphase oil	MIC	Pitting at BOL	• Fluid velocity • Operating/reservoir temperature • Nutrient contents • Pipeline cleanliness • Restricted flows (dead legs)	Rules of thumb
Injection	Produced water, com-mingled water, produced gas, WAG etc.	$H_2S/CO_2/O_2$ corrosion	Metal loss (general and/ or localised corrosion)	• CO_2 • H_2S • O_2 • Operating conditions • Water chemistry	No proven model
	Sea, brackish, aquifer or river waters, WAG, commingled water	O_2 corrosion	Metal loss (general and/ or localised corrosion)	• O_2 • Operating conditions • Water chemistry	Many models, operating company preference
	Produced water, commin-gled water, produced gas, etc.	MIC	Pitting at BOL	• Fluid velocity • Operating/reservoir temperature • Nutrient contents • Pipeline cleanliness • Restricted flows (dead legs)	Rules of thumb

BOL, bottom of line; WAG, water alternating gas.

3.7.3   O$_2$ corrosion evaluation

While several models have been developed to assess O$_2$ corrosion in water injection systems, again, there is no international standard available as indicated in Table 3.6. In general, the intention is to remove or minimise oxygen presence and to evaluative the choice of materials in respect of residual oxygen.

It should be emphasised that for H$_2$S containing environments, the presence of oxygen can lead to formation of elemental sulphur which in turn can result in elemental sulphur corrosion. The likelihood of such threat needs careful assessment.

Finally, dissolved oxygen can be fairly quickly consumed by C-steels, although this is unlikely within the limited length available on topside piping. This implies that oxygen corrosion can be excessive at the water injection pipeline inlet and not relevant at its outlet. Replacing the pipeline inlet with an appropriate corrosion-resistant option would transfer oxygen downstream and is not an option.

3.7.4   External corrosion assessment

There is no specific corrosion model to evaluate or predict the magnitude of external corrosion threats. However, the comments made in Section 3.3.4 cover the manner by which the threat is assessed.

3.7.5   Corrosion assessment for other threats

Other threats covered in Section 3.2 are system-specific and should be addressed on a case-by-case basis.

One such threat that should be considered is TOLC. In this case, if a line is only partially filled and the flow regime is stratified with the pipe wall temperature below the dew point, condensation of water can occur at the top of the line and corrosion can take place. In such situations, corrosion inhibition will not afford protection as it cannot reach the top of the line readily [7]. Specific attention is needed to assess potential corrosivity TOLC. TOLC rate can be estimated from the condensation rate, water chemistry and gas content, although there is no standard approach.

3.7.6   Overview of corrosivity assessment

From an engineering point of view, the knowledge of corrosion severity presents a convenient means of describing or classifying corrosive conditions as a basis for managing corrosion in pipelines. In addition, this knowledge aids materials selection and specification both for the new build as well as existing pipelines. Any prediction made is not only influenced by the validity of the model but also the quality and reliability of the supplied production and operational data used to calculate the corrosion risk.

There is a shift to using a more pragmatic approach to corrosion prediction moving away from the emphasis on 'corrosion rate' to 'risk categorisation', using a mechanistic approach to ensure pipeline integrity over the design life.

In PCM, consideration of positive and ongoing interaction between the corrosion engineer, production engineer/chemist, inspection engineer and pipeline engineer is essential to ensure that the relevant service conditions are defined and detailed. There has to be a common understanding of what is required against the limitations of the predictive model and subsequent monitoring/inspection. Rationalisation and integration of monitoring data, inspection information and historical trends in corrosion patterns with predicted rates are vital steps in PCM.

In the majority of cases, much emphasis is placed on the use of predicted corrosion rates at the design stage of a project. In contrast, once production starts, emphasis normally placed on measuring corrosion rates and prediction is largely replaced by trend analysis. However, in PCM, more consideration should be given to prediction/corrosion risk analysis to complement monitoring and inspection data.

Bibliography

Hotspots
- ISO 21457, 'Petroleum, petrochemical and natural gas industries – Materials selection and corrosion control for oil and gas production systems', ISO 21457: 2010, ISO, Geneva, Switzerland, 2010.
- ISO 13623, 'Petroleum and natural gas industries – Pipeline transportation systems', ISO 13623:2009, ISO, Geneva, Switzerland, 2009.
- Dt Norske Veritas, DNV OS-F101, 'Submarine pipeline systems', Offshore Standard, 2010.

Training bodies
- NACE International, USA.
- Frosio, Norway.
- Institute of Corrosion, UK.
- Afnor Competance, France.
- SPE, Worldwide.
- Different Universities and Private Organisations, Worldwide.

Corrosion threats
- 'Damage mechanisms affecting fixed equipment in the refining industry', API RP 571, API Publishing Services, Washington, DC, 2003.
- 'External corrosion awareness handbook: A guide for visual recognition of external integrity threats to upstream oil and gas production plant', Energy Institute, 2010.
- 'Corrosion threat handbook, upstream oil and gas production plant', Energy Institute, London, 2008.
- 'Managing system integrity of gas pipelines', ANSI/ASME B31.8s – 2010, ASME, 2010.
- 'External stress corrosion cracking of underground pipelines', NACE International Publication 35103, NACE International, 2003.

H_2S corrosion
- 'Petroleum and natural gas industries – Materials for use in H_2S containing environments in oil and gas production', ISO 15156:2009, Parts 1–3, ISO, Geneva, Switzerland, 2009.
- 'Guidelines on materials requirements for carbon and low alloy steels for H_2S-containing environments in oil and gas production', EFC Publication 16, The Institute of Materials, Maney Publishing, 2009.
- 'Corrosion resistant alloys for oil and gas production: Guidance on general requirements and test methods for H_2S service', EFC Publication 17, The Institute of Materials, Maney Publishing, 2002.

CO₂ corrosion

- 'Predicting CO_2 corrosion in the oil and gas industry', (eds M. B. Kermani and L. M. Smith), EFC Publication 13, The Institute of Materials, London, 1995.
- 'CO_2 corrosion control in oil and gas production – design considerations', (eds M. B. Kermani and L. M. Smith), EFC Publication 23, The Institute of Materials, London, 1997.

Soil resistivity

- 'Standard test method for field measurement of soil resistivity using the Wenner four-electrode method', ASTM G57-06, ASTM, 2006.

Mitigation

- 'Paints and varnishes. Corrosion protection of steel structures by protective paint systems', ISO 12944-5:2007, ISO, Geneva, 2007.
- 'Petroleum and natural gas industries – cathodic protection of pipeline transportation systems, Parts 1 and 2', ISO 15589, ISO, Geneva, 2004.

Corrosion under insulation

- 'Corrosion under insulation (CUI) guidelines', (ed. S. Winnik), EFC Publication 55, Woodhead Publishing, 2008.

References

1. 'Corrosion protection of steel structures by protective paint systems', ISO 12944, ISO, Geneva, Switzerland, 1998.
2. 'Recommended practice for analysis of oilfield waters', API RP 45, API, 1998.
3. B. Kermani and A. Morshed: *Corrosion*, 2003, **59**, 659–683.
4. R Nyborg: Corrosion 2010, NACE Annual Corrosion Conference, San Antonio, TX, 14–18 March 2010, NACE International, Paper 10371.
5. M. B. Kermani, J. W. Martin and K. Esaklul: Corrosion 2006, NACE Annual Corrosion Conference, San Diego, CA, 12–16 March 2006, NACE International, Paper 06121.
6. M. Bonis: 2009, NACE Annual Corrosion Conference, Atlanta, GA, 22–26 March 2009, NACE International, Paper 09564.
7. Y. Gunaltun et al.: Corrosion 2010, NACE Annual Corrosion Conference, San Antonio, TX, 14–18 March 2010, NACE International, Paper 10093.

Corrosion mitigation aims at controlling the rate and/or distribution of corrosion to a level that is economically justifiable in the context of pipeline planned life, whilst minimising human and environmental risks.

Corrosion mitigation in pipelines falls broadly into four themes:

i.   Change materials of construction through selection of appropriate materials or using an adequate corrosion allowance with regard to system corrosivity.
ii. Change/modify the operational parameters such as flow, temperature, water removal.
iii. Change/modify the fluid chemistry through removal/reduction of water, corrosive gasses (O_2, CO_2 and H_2S), addition of corrosion inhibitor or neutralising the pH.
iv. Change/modify the interfacial condition of metal through surface modification such as coating or lining and cathodic protection (CP).

Pipeline corrosion mitigation is achieved through one or a combination of these methods. The relative merits of each approach must be viewed in the context of the application, the required service life, the severity of the conditions and economy. No hard and fast rules exist in a general sense, and many decisions are made based on past experience and individual preference. Finally, the value and importance of corrosion monitoring (Chapter 5) and inspection (Chapter 6) should not be forgotten and go hand-in-hand with the above for the most part and certainly where carbon and low alloy steels are used.

This chapter (Fig. 4.1) deals with the methods, approaches and procedures in mitigating both internal and external corrosion in pipelines. It describes the implementation methods and goes on to outline the merits of each technique and their specific requirements. Finally, a section is assigned to outlining pipeline corrosion management (PCM) targets as an integrated part of the overall PCM programme.

4.1 Introduction

The mitigation of corrosion is key to maintaining the integrity of a pipeline to both achieve its design life and where required allow life extension. The mitigation process is based on slowing or stopping corrosion in the pipeline. The discussion in this section focuses on the options available to place barriers in the way to minimise corrosion in a pipeline.

The decision on what mitigation method to use starts at the concept stage of a project.

For the choice of correct mitigation method, corrosion threat needs to be understood. In the choice of mitigation method, a number of considerations should be made including:

- CAPEX and OPEX optimisation
- the risk associated with corrosion threat
- logistics (onshore or offshore)
- resource availability
- others.

The choices made at the design stage are normally based on a fixed life of the system. However, in many circumstances, the life of the system may need to be extended. For further details see Chapter 8.

The mitigation methods and associated monitoring and inspection methods are outlined in Table 4.1.

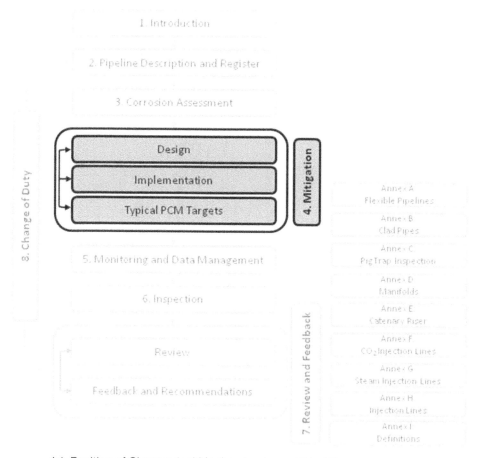

4.1 Position of Chapter 4 within the structure of this RP

4.2   Materials selection

The first step in corrosion mitigation is to determine the feasibility of using C-steel for the application. Such an assessment is described in Chapter 3. In the event where C-steel plus additional mitigation measures outlined in this chapter proves inappropriate for the application, corrosion resistant alloy (CRA) or non-metallic options can be considered. These options are covered in Annexes A and B.

Table 4.1 Examples of mitigation action grouped together under various areas and showing associated monitoring and inspection activities

Group	Mitigation method	Compliance checks (monitoring)	Verification (inspection)	Application (product line or injection line)	Examples of typical PCM targets
Materials selection	Corrosion-resistant materials	Operating envelope	No verification under normal circumstances, other than assuring operating limits are adhered to	Both	Water and gas analysis and compliance with design basis
Process control	Dehydration	Process control	Dew point/BS&W	Product	10 °C above dew point
	Operational pigging	Pigging frequency & type	Pigging records	Both	Compliance with operational pigging frequency and pig type
	Mechanical gas stripping	O_2, CO_2 and H_2S monitoring	O_2, CO_2 and H_2S concentrations	Both as applicable	Daily gas analysis reports showing compliance with design
Chemical treatments	Corrosion inhibition	Injection rates/residual inhibitor level	Fe/Mn counts Wall thickness measurements Corrosion coupons & probe data	Both	Injection rate measurements according to specification, Cl residual in water above the specified threshold value
	Oxygen scavenger (Ox Sca)	Injection rates	Residual oxygen concentration/ wall thickness measurements	Injection	Injection rate measurements according to specification, Oxygen Scavenger measured residuals below specified critical value
	H_2S scavenger	Injection rates	H_2S concentration	Both	Injection rate measurements according to specification, H_2S content below specified critical value
	Biocide	Injection rates	Biological activity	Both	Frequency of biocide treatment according to specification; sampling frequency and measured bacterial activities within the specified limits
	pH control	Injection rates	pH	Both	Injection rate and in-situ pH above the specified threshold values

Table 4.1 Continued

Group	Mitigation method	Compliance checks (monitoring)	Verification (inspection)	Application (product line or injection line)	Examples of typical PCM targets
Organic coatings	Coating and painting	Coating degradation	Coating survey/thickness measurement	Both	Periodic assessment of coating conditions by visual comparison methods using applicable standards
Cathodic protection	Impressed current	Pipeline potential TR outputs	• Potential measurements • Voltage and current outputs from TR units	Both	Frequency according to design basis and potential value in compliance with criteria from applicable standards
	Sacrificial anodes	• Pipeline potential • Anode conditions	• Potential measurements • Anode consumption	Both	Frequency according to design basis and potential value in compliance with criteria from applicable standards

One additional measure in materials selection is the use of corrosion allowance (CA). This is considered a wastage factor which is used to cater for deviations or failure to achieve PCM targets. Local legislative regulations may require a predefined CA as an integral part of the pipeline system. In such cases, the requirement takes precedence.

Particular considerations should be given to hotspots (Chapter 3) as the corrosive environment in this area can be more aggressive than other areas of the line. In particular, for top of line corrosion (TOLC) mitigation, measures need to be considered which will either avoid the condensation of water in C-steel pipelines (use of CRA cooling spools or insulation) or chemical treatment by batch treatment or spray pigging.

4.3   Corrosion resistant alloy (CRA) clad and non-metallic liners

Specific recommendations for solid/clad CRA and non-metallic liners as corrosion mitigation measures are covered in Annex B.

4.3.1   Internal coatings

In addition to non-metallic linings described in Annex B, there are coating systems that can be applied to the internal surfaces of pipelines. These coatings can be applied for a number of reasons and include flow assurance. A positive side effect of these coatings is that they produce a barrier which will mitigate internal corrosion, apart from the field joints where bare surfaces may exist.

In certain circumstances such as water injection lines, coating systems are chosen for corrosion control and also to reduce the quantity of the solid corrosion product. Accumulation of such solid products may lead to under-deposit corrosion and/or impact well injectivity.

In each case, the correct application procedure for the coating is recommended – this is dealt with in more detail in Section 4.8 on external coatings.

4.4   Chemical and physical mitigation methods

Removal of a corrosive species or rendering it ineffective can aid corrosion mitigation. This is normally achieved by physical or chemical methods, respectively, as described in this section.

4.4.1   Dehydration

Removal of water associated with hydrocarbon can reduce the corrosion threat. This is normally achieved by mechanical separation for liquids and/or by dehydration for gases. For gas lines, the gas should be dehydrated to a level so that the dew point is below the lowest anticipated temperature in the line.

Dry gas is not corrosive provided the dew point specification is adequate and adhered to. However, in some cases, incorrect conditions can cause water to be introduced in which case a review of corrosion threats should be carried out as described in Chapter 7.

4.4.2   Oxygen removal

Oxygen removal is a typical method for mitigating corrosion in water injection systems. Physical methods include vacuum deaeration and gas stripping. This operation can remove the oxygen down to levels of a few tens of parts per billion (ppb). This can be supplemented by a chemical method of oxygen removal as described in Section 4.4.4. If incorrect conditions cause oxygen to be introduced, a review of corrosion threats should be carried out as described in Chapter 7.

4.4.3   Chemical treatment injection methods

The majority of the internal corrosion mitigation methods involve the use of chemicals. An important consideration in the choice and deployment of such chemicals is their interaction and compatibility which may require special considerations.

Corrosion inhibitors: This is probably the most common form of internal corrosion mitigation used in the oil and gas industry. The method involves the injection of a corrosion inhibitor at a dose in the range of 10 to 1000 parts per million (ppm) continuously or a batch treatment of inhibitor at 1% to 20%. The type and amount of corrosion inhibitor required depend upon system corrosivity.

   The use of a corrosion inhibitor can achieve a corrosion rate typically less than 0.1 mm year^{-1} in most cases.

   The selection process for the inhibitor (including laboratory assessment) should reflect pipeline operating conditions and status and should not look into general corrosion alone [1].

4.4.4   Scavenging chemicals

There are two main types of scavenging chemicals used for corrosion control – these are oxygen scavengers and hydrogen sulphide scavengers.

Oxygen scavengers: These are continually injected to reduce dissolved oxygen to acceptable levels to supplement physical deaeration processes.

   Use of oxygen scavengers should recognise that it may interact with biocides and chlorination.

Hydrogen sulphide scavengers: These are used for mitigating some forms of H_2S corrosion threat. The scavengers are injected continually or as a batch chemical depending on the application. The chemicals are best used in gas systems but can equally be used in multiphase systems with lower efficiency.

4.4.5   Scale inhibitors

While not directly affecting the corrosion process, scale inhibitors have a role to play in corrosion mitigation as they can eliminate an environment where corrosion may readily occur by controlling solids deposition.

4.4.6   Biocides

These are deployed to mitigate microbial induced corrosion (MIC).

   There are two types of biocide: (i) oxidising agents (such as chlorine) and (ii) organic chemicals (such as glutaraldehyde). Typically, the former is used in aerobic systems while the latter is used for anaerobic systems such as with sulphate-reducing bacteria (SRB). This is an area where the compatibility of chemicals is paramount.

Chlorine is a corrosive species in itself and therefore residual chlorine in the system needs to be controlled.

Organic products are generally applied as batch treatments. It is common practice for more than one type of biocide to be alternated during the treatments to avoid the bacteria flora becoming immune to one type.

4.4.7   Multi-component chemicals

Multi-component chemicals such as those used for hydrotesting (i.e. biocide + oxygen scavenger + corrosion inhibitor), or production fluids (i.e. corrosion + scale inhibitor) should be evaluated for their efficacy.

4.5   Operational pigging

Pigging aids corrosion mitigation by removing water or solids from the line and is also recommended to improve deployment of batch treatment corrosion inhibitors or biocides. The chemical and microbiological characterisation of solids removed by pigs is helpful in defining the corrosion threat.

It should be noted in the case of lined or internally coated pipes that the use of operational pigging is not recommended as the brushes may cause damage to the coatings thereby affecting its integrity.

4.6   Top of line corrosion (TOLC) and control

Mitigating TOLC can be achieved by several methods including:

- dehydration
- thermal insulation of pipe
- CRA cooling spool
- corrosion allowance to manage high potential corrosion rates during the early life of the line
- periodic batch treatment with a concentrated inhibitor or spray pigging (i.e. spraying the inhibitor from the bottom to the top of the pipe)
- neutralisation of condensed water by volatile organic inhibitors (VOC) – this is a design solution which is not normally acceptable, but may be employed during the operation
- neutralisation of organic acids by pH stabilisation (partial mitigation, only reducing the contribution of organic acids).

4.7   pH stabilisation

The key principle of pH stabilisation is to favour the precipitation of a protective corrosion product on the surface (iron carbonate in sweet environments and iron sulphide in sour conditions). A sufficient pH must be reached to guarantee the complete protectivity. There are two limitations to pH-stabilisations including: (i) water chemistry (presence of salts leading to formation of mineral scales) and (ii) the overall acidity of the effluent which may require an unrealistic quantity of pH-stabiliser – further details are given in the Bibliography (Section 4.13 and ISO 21457).

In contrast, the main advantages of pH-stabilisation include: (i) avoiding the production of solids from residual corrosion products usually experienced with corrosion inhibitors thereby suppressing the need to handle these solids at the end of the line, (ii) using chemicals with less impact on the environment than corrosion inhibitors.

This solution involves injection of an alkaline product (either organic as methyl-di-ethanol-amine (MDEA) or mineral as NaOH or KOH) to the system, in a sufficient concentration to increase the pH of the condensed water above a sufficient value to guarantee that this water is no longer corrosive for CS pipelines. This is normally performed in association with glycol injection used for hydrate control.

This pH stabiliser is added in a large concentration to the system and is normally recycled with re-generated glycol through returned lines. pH stabiliser top-up to maintain the target pH is usually made once to twice a year, and even less in some cases. Larger top-ups may, however, be required to the agreed target after pig cleaning operations. This is required to re-establish the pH target in the freshly condensed water. Frequent sampling, water analysis and pH measurements of the outlet fluids are used to confirm the condition of the fluid and that the target pH has been reached.

4.8   External coatings

The most widely used method for controlling external corrosion is the use of suitable barrier coatings applied, maintained and monitored in the correct manner. Normally, coating systems are used as the primary barrier to provide protection of above ground lines. However, such systems are used in association with CP systems for protecting buried or submarine lines.

The coating system can be made up of a number of layers but the correct selection of the coating system, required surface preparation and proper application of all of the possible coats is key to its successful application. For detailed procedures and requirements for surface preparation, coating selection and application, refer to the Bibliography.

Particular attention should also be made to coating being applied to the splash zone, risers and land falls. In these areas, the corrosive environment is particularly harsh and a high performance coating such as a neoprene coating should be employed.

The coatings can be shop applied before the line pipe is supplied or applied in situ. The latter is used to coat areas where welding takes place or where repairs are required if the shop applied coating gets damaged. Further guidance on coating applications can be found in international documents outlined in Section 4.13.

It should be noted that coating of landfall is not covered in existing standards – therefore, the following points need particular attention during landfall design:

- Requirement for high quality external pipeline coating (selection and application).
- Philosophy for avoidance of damage to the coating during installation, assurance and repair.
- Provision of electrical insulation of pipeline from casings, pipe-supports, bellmouths, armouring in concrete structures, etc.
- Provision of required sampling, inspection and monitoring of the coating and surrounding environments.

4.9   Cathodic protection (CP)

CP is used to mitigate external corrosion on buried or submarine pipelines. This can be achieved in two ways: (i) sacrificial anodes or (ii) impressed current systems.

Detailed requirements, information and implementation methods for CP systems and for the design, installation and monitoring of these systems are covered extensively in respective publications outlined in the Bibliography.

Particular attention should be paid to systems where sacrificial anodes and impressed current systems are used in combination. This may be the case at landfalls or close to industrial installations. In addition, care should be exercised to cater for the interference caused by stray current from high voltage power lines or telluric current. CP (intentional or otherwise) of CRA also needs specific care as described in Annex B.

4.10   Competency

With all these processes, the competency of personnel selecting mitigation methods, designing the system or installing and commissioning should be assessed by project specific qualification schemes. While competency can be assessed by company or project specific qualification schemes, there are tailored courses and certification available for mitigation methods including:

- Coating inspector
- Coating applicator
- CP Engineer levels
- Internal pipeline corrosion
- Chemical treatment specialist
- Internal corrosion control.

A level of competency against various courses should be defined for all personnel working on a project from design through installation to operation. Further details are covered in Chapters 2 and 3.

4.11   Overview

To summarise the many mitigation methods possible for corrosion management, a matrix can be drawn up. Within this matrix, the monitoring and inspection methods associated with the mitigation process can also be defined.

The example matrix in Table 4.1 shows various mitigation methods and groups them under various sections with reference to compliance checks and verification.

4.12   Setting PCM targets

To ensure intended mitigation actions are implemented, appropriate targets should be assigned and reviewed frequently as described in Chapter 7. Depending upon the pipeline system, the targets may differ. However, these will be linked not only to the mitigation methods described in the present chapter, but also to the monitoring and surveillance/inspection activities discussed in Chapters 5 and 6.

Common key targets should address the following typical considerations:

- Is the corrosion rate, including localised damage rate, within the acceptable range set for the pipeline?
- Is the efficiency of mitigation systems being met?
- Are corrosion monitoring systems functioning as intended (including frequency of sampling and analysis) and respective data within the acceptable limits?
- Is the pipeline subject to In-line Inspection (ILI) per the PCM programme and respective data within the acceptable range set for the pipeline?
- Are data arising from any other pipeline inspection activities or operational pigging in line with pipeline PCM targets?
- Are the nature and composition of the transported fluids in line with the intended design?
- Is the quality of injected fluids within the expected range, i.e. dissolved O_2 content of injection water?

Examples of typical targets are given in Table 4.1.

As part of the review and feedback process described in Chapter 7, where service changes occur or in response to monitoring/inspection data, PCM targets may require adjustments during the lifetime of the pipeline. Failure to meet the PCM targets would require full fitness for service assessment to determine the consequences of deviation on pipeline status. This should include a full review of PCM as discussed in Chapters 7 and 8.

Bibliography

Internal lining or cladding
- API 5L2, Recommended Practice for Internal Coating of Line Pipe for Non-Corrosive Gas Transmission Services.
- API 5L7, Recommended Practices for Unprimed Internal Fusion Bonded Epoxy Coating of Line Pipe.
- NACE RP-191-2002, The Application of Internal Plastic Coating for Oilfield Tubular Goods and Accessories.
- NACE SP-0181-2006, Liquid – Applied Internal Protective Coatings for Oilfield Production Equipment.
- BS10339:2007, Steel Tubes for Onshore and Offshore Water Pipelines – Internal Liquid Applied Epoxy Linings for Corrosion Protection.
- ASTMG48, Clad Piping Testing Requirements.
- BS10339:2007, Steel Tubes for Onshore and Offshore Water Pipelines – Internal Liquid Applied Epoxy Linings for Corrosion Protection.

Availability of corrosion inhibitor
- NORSOK M-001, Materials Selection Standard.

External coating standards
- DNV-RP-F102, Coating Specification for Field Joint and Linepipe Coatings.
- DNV-RP-F106, Factory Applied External Pipeline Coatings for Corrosion Control.
- API RP 5L9, External Fusion Bonded Epoxy Coating of Line Pipe.
- BS EN ISO 21809-2, Petroleum and Natural Gas Industries – External Coatings for Buried or Submerged Pipelines Used in Pipeline Transportation Systems.
- ISO 8501-1, Preparation of Steel Substrate before Application of Paints and Related Products – Visual Assessment of Surface Cleanliness.

- ISO 11124-1, Preparation of Steel Substrate before Application of Paints and Related Products – Specifications for Metallic Blast-cleaning Abrasives.
- BS EN ISO 21809-2, Petroleum and Natural Gas Industries – External Coatings for Buried or Submerged Pipelines Used in Pipeline Transportation Systems.
- API 5L2, Recommended Practice for Internal Coating of Line Pipe for Non-Corrosive Gas Transmission Services.
- API5L7, Recommended Practices for Unprimed Internal Fusion Bonded Epoxy Coating of Line Pipe.
- RP-191-2002, The Application of Internal Plastic Coating for Oilfield Tubular Goods and Accessories.
- ISO 12944, Part-5, Paints and Varnishes – Corrosion Protection of Steel Structures by Protective Paint Systems.
- ISO 20340, Paints and Varnishes – Performance Requirements for Protective Paint Systems for Offshore and Related Structures.
- ISO 21809, Petroleum and Natural Gas Industries, External Coatings for Buried or Submersed Pipelines Used in Pipeline Transportation Systems.
- ISO 8501, Parts, 1 to 5, Preparation of Steel Substrates before Application of Paints and Related Products – Visual Assessment of Surface Cleanliness.
- ISO 8502, Preparation of Steel Substrates before Application of Paints and Related Products – Tests for the Assessment of Surface Cleanliness.
- NACE Publication 10D199.
- NACE Technical Committee Report Item 24202, Internal Corrosion Monitoring of Subsea Production and Injection Systems, 1D199.
- NACE RP 0394, Standard Recommended Practice – Application, Performance, and Quality Control of Plant-Applied, Fusion-Bonded Epoxy External Pipe Coating.
- NACE RP 0399, Plant-Applied, External Coal Tar Enamel Pipe Coating Systems: Application, Performance, and Quality Control.
- DIN 30670, Polyethylene Coatings of Steel Pipes and Fittings; Requirements and Testing.
- DIN 30678, Polypropylene Coatings for Steel Pipes.
- DNV-RP F106, Factory Applied External Pipeline Coatings for Corrosion Control.
- DNV-RP-F102, Pipeline Field Joint Coating and Field Repair of Linepipe Coating.
- NACE RP 0402, Field-Applied Fusion-Bonded Epoxy (FBE) Pipe Coating Systems for Girth Weld Joints: Application, Performance, and Quality Control.
- NACE RP 0602, Field-Applied Coal Tar Enamel Pipe Coating Systems: Application, Performance, and Quality Control.
- NACE RP 0303, Field-Applied Heat-Shrinkable Sleeves for Pipelines: Application, Performance, and Quality Control.
- NACE RP 0105, Liquid-Epoxy Coatings for External Repair, Rehabilitation, and Weld Joints on Buried Steel Pipelines.

Sacrificial anode CP systems
- DNV-RP-B401, Cathodic Protection Design.
- DNV-RP-F101, Corroded Pipelines.
- ISO 15589-2, Petroleum and Natural Gas Industries – Cathodic Protection of Pipeline Transportation Systems – Part 2: Offshore Pipelines.

- DNV-RP-F103, Cathodic Protection of Submarine Pipelines by Sacrificial Anodes.
- ISO 15589-1, Petroleum and Natural Gas Industries – Cathodic Protection of Pipeline Transportation Systems – Part 1: On-land Pipelines.
- BS7361-1, Code of Practice for Land and Marine Applications.

Impressed current CP systems
- ISO 15589-1, Petroleum and Natural Gas Industries – Cathodic Protection of Pipeline Transportation Systems – Part 1: On-land Pipelines.
- BS7361-11991, Cathodic Protection – Part 1, Code of Practice for Land and Marine Applications – (formerly CP 1021).
- NACE SP0572-2007, Design, Installation, Operation and Maintenance of Impressed Current Deep Anode Beds.
- NACE SP0169-2007, Control of External Corrosion on Underground or Submerged Metallic Piping Systems.
- ISO 15589, Petroleum and Natural Gas Industries – Cathodic Protection of Pipeline Transportation Systems – Part 1: On-land Pipelines; Part 2: Offshore Pipelines.

CP design
- DNV-OS-F101, Submarine Pipeline Systems.
- ISO3183, Petroleum and Natural Gas Industries – Steel Pipe for Pipeline Transportation Systems.
- DNV-RP-F105, Free Spanning Pipelines.
- EN 12954, Cathodic Protection of Buried or Immersed Metallic Structures – General Principle and Application for Pipelines.
- NACE SP0572, Design, Installation, Operation and Maintenance of Impressed Current Deep Anode Beds.
- NACE SP0169, Control of External Corrosion on Underground or Submerged Metallic Piping Systems.
- NACE SP 0286, Electrical Isolation of Cathodically Protected Pipelines.

Materials selection
- ISO 21457, Petroleum, Petrochemical and Natural Gas Industries – Materials Selection and Corrosion Control for Oil and Gas Production Systems.
- NORSOK M-001, Materials Selection Standard.

Reference

1. J. W. Palmer, W. Hedges and J. L. Dawson (eds): 'Use of corrosion inhibitors in oil and gas production', No. 39 in European Federation of Corrosion Series. Maney, London, UK, 2004.

Corrosion monitoring and data management

Corrosion monitoring may be defined as the systematic and continual direct or indirect measurement of system corrosivity. Its objective is to assist in the understanding of the corrosion process and/or to obtain information for use in controlling corrosion and reviewing mitigation actions. This chapter (Fig. 5.1) outlines the requirements, methods and information required to carry out internal as well as external corrosion monitoring of pipelines. It includes a section on data management as successful pipeline corrosion management (PCM) is entirely dependent on the quality of information gathered and its respective analysis and interpretation.

It is emphasised that due to the localised nature of all monitoring techniques, regular inspection by intelligent pigging remains the sole means of obtaining full information on corrosion damage and its intensity over the entire length of the line.

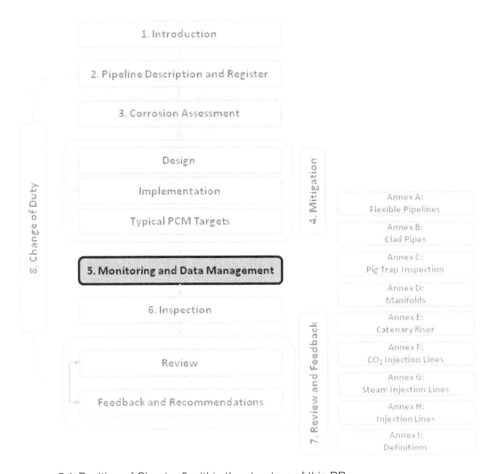

5.1 Position of Chapter 5 within the structure of this RP

5.1   Introduction

It is of paramount importance to monitor corrosion before and after any changes in service conditions and/or mitigation measures. The knowledge gained from an effective monitoring system allows the characterisation of fluid corrosivity and the performance of mitigation actions. The information gained may also assist in the optimisation of similar systems for future projects. Direct or indirect corrosion monitoring can be challenging and reliance should not be placed on a single method or data from a single location whenever operationally possible.

In this RP, direct and indirect corrosion monitoring methods are defined according to NACE International Publication 3T199.

5.2   Internal corrosion monitoring

Continuous or periodic pipeline monitoring activities may be divided into several categories, including:

- fluids and process monitoring
- corrosion monitoring (e.g. local corrosion measurements).

Results from monitoring are important inputs to the PCM programme and can be used to define the need for inspection. The extent of monitoring should be defined in the overall PCM programme and will be a function of pipeline criticality, pipeline material and corrosivity of transported fluids.

It should be noted that in the context of this RP, fluids include gas, water and liquid hydrocarbon.

5.2.1   Fluids and process monitoring

Fluids and process monitoring provides information about the parameters that may cause, affect or remediate internal corrosion. Such parameters should be identified for each pipeline and should include the monitoring of chemical and physical mitigation methods. The programme should reflect the critical parameters that have been identified in the corrosion assessment (Chapter 3). Typical elements of fluid and process monitoring include some or all those outlined in Table 5.1. As an example, for many dry gas pipelines, only water dew point need be monitored.

Fluid parameters are normally obtained by sampling, e.g. water samples or gas samples. Process parameters can normally be obtained by continuous monitoring. Solid characteristics may be obtained from pigging samples and therefore the pigging debris should be assessed after each pigging operation.

Parameters which can vary from the inlet to the outlet of the pipeline should be monitored at both locations if operationally feasible.

Based on these parameters, potential corrosion or system corrosivity over time can be estimated from relevant corrosion models as discussed in Chapter 3. Depending on the type of system, some parameters can be used to identify whether a specific corrosion threat is relevant. An example includes increased H_2S at the outlet of a pipeline which may indicate bacterial activity.

For pipeline systems with pH stabilisation, fluid analyses of rich and lean monoethylene glycol (MEG) will be a part of the regular monitoring of the system, with particular focus on the concentration of bicarbonate and/or pH stabiliser and the in-situ pH.

Table 5.1 Typical elements of internal corrosion monitoring

System and intent	Parameters, methods and criteria	
Fluids, solids and process monitoring	Water content CO_2 content H_2S content Oxygen content Elemental sulphur Mercury Water compositions (anions and cations, alkalinity, organic acids) Iron counts (both as ions in the water and as solid particles), manganese counts Residual amounts of injected corrosion inhibitors Temperature Pressure Flow rate and flow regime Chemical injection rates Water dew point of gas Production rate of solids, sands, scale, proppant Operational changes Solids composition Bacterial monitoring (water sampling and bacterial counts with test-kits)	
Corrosion monitoring	Corrosion monitoring; indirect methods	Commonly used coupons and probes include: • Weight loss coupons (WL) • Electrical resistance (ER) probes • Linear polarisation resistance (LPR) probes • Bio probes
	Corrosion monitoring; direct methods	Available methods for direct measurements include: • Permanent online measurement equipment (mostly ultrasonic type techniques); permanent ultrasonic methods (UT mats, medium range guided waves)
Physical and chemical treatments	Verification of injection rates	• Instantaneous injection rate measurement (manual or automatic) • Daily level control of tanks • Weekly measurement of stock + supply on each injection unit • Monthly control of quantities delivered (global evaluation per field or site/platform) • Percentage uptime for chemical injection
	Verification of chemical dose rates	• Dosage of active chemicals (inhibitor, pH, alkalinity, phosphates) • Dosage of reaction products (residual O_2, residual chlorine, pH, residual sulphite)
	Verification of physical treatments	• Water dew point on dry gas • CO_2 + H_2S content downstream gas removal units • Efficiency of vacuum stripping towers for water de-aeration • Operational changes
	Treatment efficiency	• Corrosion (weight loss) coupons • ER probes • LPR, highly sensitive ER probes • Dosage of iron content
	Mixed inspection monitoring tools	• External permanent UT monitoring

5.2.2   Corrosion monitoring

Corrosion monitoring is carried out using a number of measurement techniques as outlined in Table 5.1. In choosing a corrosion monitoring technique and location, their relevance to the applicable corrosion threat should be carefully evaluated.

The choice of monitoring technique should be made at an early stage of the design process to ensure that it can be implemented. Furthermore, it is critical that any modifications required during construction are fully assessed to ensure that the design intent is met.

After construction, relevant equipment should be fully commissioned by competent personnel to ensure their reliable operation once production starts.

Further details on direct and indirect corrosion monitoring techniques can be found in the Bibliography (Section 5.5).

For systems protected by corrosion inhibitors, results from monitoring may be used to ensure adequate injection rates. This may require a high sensitivity and fast response time, preferably on a continuous basis.

For dry gas pipelines operating at a temperature at least 10 °C above the dew point, internal corrosion monitoring systems may not be required and the most appropriate monitoring is process monitoring for moisture excursions.

Typical selection and application of different corrosion monitoring techniques are shown in Table 5.2.

5.2.3   Unpiggable lines

It should be noted that corrosion monitoring is not a substitute for inspection. However, in instances where in-line inspection of carbon steel flowlines is impractical or impossible, then the risk-based assessment of such lines can be better informed where corrosion monitoring is present at the inlet and outlet.

5.2.4   Observation from specific operations

Pertinent information should be collated when specific operations are performed. Such operations include but are not limited to:

- Operational pigging: parameters include solids, bacterial count, water analysis, etc
- Start up: water analysis, bacterial count, H_2S content, calcium and other salt content, iron count, etc
- Acid cleaning: in-situ pH, calcium and other salt content, etc
- Hydrotesting: preservation/moth balling; bacterial count, residual biocide and oxygen scavenger, iron count, etc.

5.3   External corrosion monitoring

Cathodic protection (CP) potential surveys are the most common methods used in the industry to monitor the external corrosion of pipelines. The pipeline industry has developed many criteria to assess CP effectiveness. These criteria may require the measurement of On or Off potentials. Therefore, the monitoring technique may be chosen depending on the selected CP criteria. Permanently installed monitoring devices for cathodically protected pipelines should be considered, especially when

Table 5.2 A general guide to the application of internal corrosion monitoring techniques

Type of service	Direct methods				Indirect methods						
	Weight loss coupons/spool piece	Electrical resistance probe	Electrochemical probes[1]	Galvanic probes	Bacterial monitoring	Solids	Dissolved gasses	pH	Hydrogen probe	Flexible UT techniques[6]	Corrosion product analysis
Water injection	✓	✓[5]	✓[4]	✓	✓	✓	O_2, Cl_2	✓		✓	✓
Crude oil, water, gas service	✓	✓	✓[3]		✓	✓	CO_2, H_2S	✓	✓	✓	✓
Aquifer water	✓	✓	✓		✓	✓	CO_2, H_2S	✓		✓	✓
Unstablised crude oil	✓	✓			✓	✓	CO_2, H_2S		✓	✓	✓
Gas and conden-sate	✓	✓					✓		✓[7]	✓	✓
Produced water	✓	✓	✓[2]		✓	✓	CO, O_2, H_2S	✓		✓	✓

[1]LPR unsuitable for hydrocarbon dominated systems and/or low conductivity fluids and/or H_2S containing fluids.
[2]May be used where oxygen content is high.
[3]Only where continuous free water phase is present.
[4]Depends on water quality. LPR unsuitable where there is biofilming tendency.
[5]Intrusive probes preferred. Flush mounted unsuitable where there is biofilming tendency.
[6]Maximum temperature 120 °C.
[7]May be useful for condensed water in sweet gas export lines.

the line is not accessible for frequent potential measurements. Such monitoring may include stab plates (offshore), reference electrode(s) for potential measurement, monitored anodes for current determination and/or coupons and probes.

When the potential survey indicates that the pipeline is under-protected, more complex actions such as the use of fault location methods and/or excavation should be carried out to assess potential coating defects. The key elements of external monitoring are included in Table 5.3.

5.3.1 Onshore

Guidance for potential surveys and related measurement techniques are included in international documents outlined in the Bibliography (Section 5.5).

After construction, CP monitoring equipment should be fully commissioned and documented by competent personnel to ensure reliable operations.

Measurement of pipeline current (line currents) may be undertaken for pipeline survey work and trending of the CP. This is particularly important for isolating joints and other locations known to pose challenges for CP systems. Examples include boreholes, landfalls, road/river crossings and the proximity of transformer rectifier

Table 5.3 Typical elements of external corrosion monitoring

System and intent	Parameters, methods and criteria	
External CP	Verification of rectifiers Efficiency of CP, onshore pipelines (protected by impressed current) Efficiency of CP, offshore pipelines (protected by sacrificial anodes)	• Output potential, current and anode resistance • Potential surveys (ON/OFF) • Corrosion coupons • Closed interval survey (CIS) • Interference monitoring (AC and DC) • Potential surveys • Anode consumption
External coating systems	Coating monitoring	• Excavation (direct assessment) • CIS, direct current voltage gradient (DCVG)

(TR) locations. Guidance on such measurements can be found in the Bibliography (Section 5.5).

The requirements for periodic surveys of the coating of onshore pipelines using CIS should be determined taking into account the selected coating and its predicted degradation, the type of soil, the observed CP potentials, the current demands and known metal loss.

The possible interference from alternating current (AC), direct current (DC) or telluric activity during the surveys and the interpretation of the results should be considered when selecting the survey method. CP surveys can be performed to provide more detailed information concerning the corrosion protection of the pipeline. Such surveys should be carried out when abnormal coating damage, severe corrosion conditions and/or stray current interferences are suspected.

5.3.2   Offshore

The general principles for monitoring of CP systems for offshore sections are given in the Bibliography (Section 5.5).

In brief, a visual inspection should be undertaken to estimate anode consumption and locate coating holidays (location and estimated size) along the pipeline. Continuous true potential measurement, continuous near line potential gradient measurements, continuous potential measurements by remote electrodes and discontinuous potential measurements on stab plates can be used to map CP potential along the pipeline. When the design of CP for offshore pipelines is highly optimised, the need for monitoring becomes increasingly important.

5.4   Data management

Following corrosion monitoring data acquisition, it is paramount that the data are stored, analysed and interpreted. The volume of acquired data should be anticipated so that the corrosion monitoring programme can be optimised accordingly taking into account this laborious and time consuming task.

Data trending should be performed and the results presented preferably in a graphical format to ease the corrosion review process (Chapter 7).

Bibliography

Related standards
- API RP45, Analysis of Oilfield Waters.
- API RP38, Biological Analysis of Subsurface Injection Waters.

Corrosion monitoring
- NACE RP 0497, Field Corrosion Evaluation Using Metallic Test Specimen.
- NACE RP 0775, Preparation, Installation, Analysis and Interpretation of Corrosion Coupons in Oilfield Operations.
- NACE TM 0194, Field Monitoring of Bacterial Growth in Oil and Gas Systems.
- NACE TM 0106, Detection, Testing and Evaluation of Microbiologically Influenced Corrosion (MIC) on External Surfaces of Buried Pipelines.
- DNV-RP-F116, Integrity Management of Submarine Pipeline Systems.
- NACE International Publication 3T199, Techniques for Monitoring Corrosion and Related Parameters in Field Applications.

CP
- NACE RP 0169; Control of External Corrosion on Underground and Submersed Subsea Systems.
- NACE RP 0104; The Use of Coupons for CP Monitoring Applications.
- NACE TM 0497; Measurement Techniques Related to Criteria for CP Underground or Submerged Metallic Piping Systems.
- ISO 15589; Petroleum and Natural Gas Industries – Cathodic Protection of Pipeline Transportation Systems – Parts 1 and 2.
- EN 12954, Cathodic Protection of Buried or Immersed Metallic Structures – General Principles and Application for Pipelines.
- BS 7361-1, Cathodic Protection; Code of Practice for Land and Marine Applications.
- EN 14545, Cathodic Protection of Complex Structures.

Coating monitoring
- DIN 30670, Polyethylene Coatings for Steel Pipes and Fittings.
- DIN 30678, Polypropylene Coatings for Steel Pipes.

6

Inspection

Corrosion inspection is a key activity in ensuring pipeline integrity. Inspection normally refers to the evaluation of the condition of a pipeline in relation to a standard or a specification. As product types and their associated requirements have become increasingly complex, inspection has also evolved necessitating involvement of different skills and state-of-the-art technologies.

Within the context of pipeline corrosion management (PCM), inspection consists of a series of steps and actions including data gathering, analysis and management outlined in the present chapter (Fig. 6.1). These are fed into corrosion assessments and reviews.

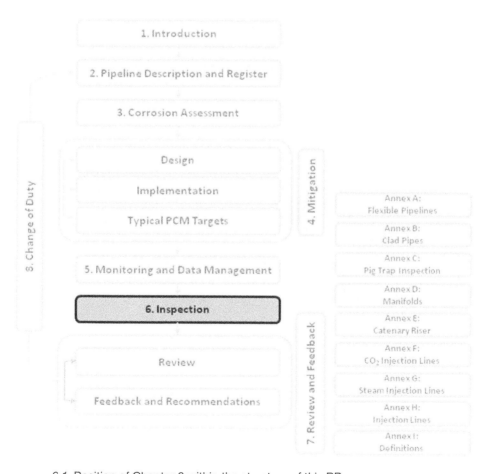

6.1 Position of Chapter 6 within the structure of this RP

6.1   Introduction and objectives

The objectives of inspection within a PCM programme are to:

- comply with local regulations and/or legislations for pipeline operation
- verify that the pipeline is fit for service under the defined operating envelope
- verify the results given by leading indicators such as corrosion monitoring data, chemical injection data and cathodic protection (CP) data
- detect internal and external corrosion defects which may not be indicated by localised corrosion monitoring sites
- verify the evolution of wall loss with time
- determine the remaining service life.

Inspection may cover the whole pipeline length by intelligent pigging, or local inspection (including specific features). Inspection is a complementary step in the overall PCM and should not be considered a substitute for corrosion monitoring.

Details of individual inspection techniques are excluded from this chapter as these are documented widely in the published literature (refer to Bibliography). However, the chapter provides guidance on the objectives of inspection, the use of risk-based inspection (RBI) and some of the key issues to be aware of in using inspection within PCM.

A key factor in inspection is to maximise defect detection (the extent of coverage and resolution of individual tools). Where corrosion occurs, it is typically due to localised attack. Therefore inspection must cover the maximum possible area and have sensitivity for small defects.

The overall process of inspection for pipelines in service is shown in Fig. 6.2.

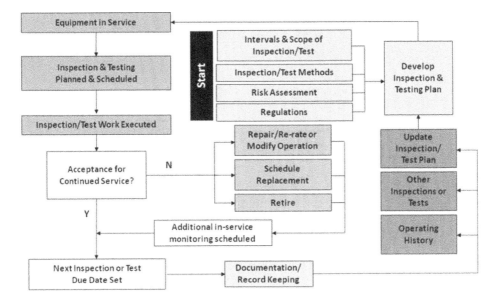

6.2 Typical inspection tasks with integrity management of pipelines

6.2   Inspection planning

The inspection plan, including techniques, frequencies and locations should be based on a full understanding of:

- the objectives of the inspection including verification of corrosion mitigation and/or detection of unmitigated corrosion threats (outlined in Chapters 3 and 4) and materials degradation
- conformance with and adequacy of PCM targets as outlined in Section 4.12.
- the materials and damage mechanisms in operation – identified through corrosion assessment (Chapter 3)
- the remaining life and mitigation requirements
- regulatory requirements
- risk assessment outputs
- historical knowledge of mechanisms of failures.

The inspection plan can include fixed scheduled or risk-based methodologies and its operating company's preference or may be subject to local regulations. There is an industry trend towards the implementation of the latter methodologies.

6.3   Baseline inspection

For any newly constructed pipeline, some operating companies may recommend to undertake a baseline inspection during or soon after commissioning. This is carried out primarily to identify manufacturing defects and to set a reference for subsequent inspection surveys. In some other cases, a further baseline inspection may be carried out to identify corrosion in pipelines which have been constructed over an extended period, or have had long lay ups before entering service – in such cases, corrosion may cease once the pipeline commences service.

6.4   Use of risk-based inspection

Risk-based inspection (RBI) schemes are planning tools developed to optimise inspection activities. RBI uses the findings from a formal risk analysis, such as a Corrosion Risk Assessment, to guide the direction and emphasis of the inspection planning and the physical inspection procedures.
  A risk-based approach to inspection planning is used to:

i.   ensure risk is reduced to as low as reasonably practicable (ALARP)
ii. optimise the inspection schedule
iii. focus the inspection effort onto the most critical areas
iv. identify and use the most appropriate methods of inspection.

In the context of PCM, RBI may be used to prioritise the inspection of pipelines within an asset or regional infrastructure, or to prioritise the inspection of individual segments or discreet locations, such as welds, within a single pipeline.
  Risk-based tools can include:

- prioritisation ranking tools based on simple criteria such as age, corrosion mitigation, failure history, inspection, etc
- fully quantitative and statistical analysis based on failure frequency, event likelihood and quantitative modelling of consequences.

RBI methodologies are well described in published documents and have been adopted by many operating companies – these are outlined in the Bibliography (Section 6.9).

6.5   Scheduled inspection

Frequency of inspection can be determined either by local regulations and legislations and/or the risk and consequence of a failure. The inspection type and interval should be established by an approved process. Therefore, frequency of inspection may vary for each pipeline.

6.6   Types of technique

The preferred method of inspection for piggable pipelines is in line inspection (ILI), also known as intelligent pigging. However, there is more than one type of inspection technique for use in ILI including MFL, UT, eddy current and EMAT. When ILI cannot be implemented (unpiggable pipelines), alternative inspection techniques should be considered for high corrosion risk areas.

The type of inspection technique should be determined by the anticipated corrosion threat. It is therefore, imperative that every effort is made to accurately predict the failure mode and the most appropriate inspection locations during corrosion assessment (Chapter 3).

The key inspection techniques available are:

- Intelligent pigging
  o   MFL
  o   UT
- Hydrotest: this can be used for confirmation of fitness for purpose, however it is only a 'single point' check, and gives no information about the extent of defects present within a pipeline. Hydrotesting, while not a corrosion inspection tool, confirms that a pipeline is fit for a given operating pressure, on a given date, but offers no further verification beyond that time.
- Direct Assessment: this is described in detail elsewhere [1]. It is a process based on an assessment of all available information and local inspection of the pipeline by UT or other methods. External corrosion direct assessment (ECDA) is focused on detection of external corrosion defects.
- Localised inspection.
- Close Interval Survey (CIS): these surveys are completed to assess coating condition. This technique will not provide any information on corrosion defects.
- Direct Current Potential Gradient Survey (DCVG): this type of survey is an above-ground survey technique, primarily used to locate external coating defects on buried pipelines. DCVG can also be used to determine the severity and orientation of defects in areas of high stray current activity.

These techniques are described widely in the literature.

Some inspection technologies, for example, Transverse MFL, UT Crack Detection, EMAT or a combination of these are likely to be used only in special circumstances where the corrosion assessment or previous inspection data have identified anomalies.

6.7   Corrosion rate determination from inspection data

One of the key challenges in interpreting inspection data is to extrapolate defect prop-agation from one inspection to another. This is particularly the case for ILI data (also applicable to other inspection techniques) where the sensitivity of the tools may be as much as several per cent of wall thickness. The location of defects is often referenced to either girth welds or distance. In both cases, it can be easy to obtain data which can give unrealistic corrosion rates. Nevertheless, such data are still of importance where no other data are available or where historical trends is the intent.

Whenever possible, verification of ILI data should be carried out with inspection measurements from outside the pipe to confirm the defect sizing accuracy of the tool.

6.8   Data management

Inspection of a pipeline results in the delivery of a standard inspection report and feature list. The contents of the report are usually specified in the contract scope of work.

Inspection reports should be assessed and acted upon within a defined timescale after the inspection to ensure that appropriate actions are derived and implemented. This may be validation of metal loss indications by follow-up inspection, or updating a corrosion risk assessment.

Bibliography

RBI
* API 581, Base Resource Document – Risk-Based Inspection.
* DNV RP G-101, Risk Based Inspection of Offshore Topsides Static Mechanical Equipment.

Direct Assessment
* NACE RP0502, Standard Recommended Practice – Pipeline External Corrosion Direct Assessment Methodology.
* NACE SP0206, Internal Corrosion Direct Assessment Methodology for Pipelines Carrying Normally Dry Natural Gas (DG-ICDA).

Reference

1. 'Standard recommended practice – Pipeline external corrosion direct assessment method-ology'. NACE RP0502-2002, National Association of Corrosion Engineers, Houston, TX, 2002.

Review and feedback are the cornerstones of an effective and successful pipeline corrosion management (PCM) programme. The purpose of such review is to evaluate PCM conduct against the defined PCM targets set out within the PCM programme. It should be emphasised that this process requires management commitment and understanding of the implementation of the PCM programme. Fully realised, review and feedback are holistic processes, bringing together many of the elements which go to make up the successful practice of a PCM programme covered in this RP. For this very reason, it is a complex process. This chapter (Fig. 7.1) outlines the necessary steps in the review and feedback process. A typical process for the review and assessment of PCM and inspection data is shown in Fig. 7.2 and falls into two broad areas as described in Sections 7.3 and 7.4.

The chapter does not go into all of the details, but it gives an overview to allow a better understanding of the fundamental steps in managing pipeline systems.

7.1   Introduction

The main objective of review is to provide productive feedback on each element of the PCM programme so that necessary improvements can be made to enhance pipeline operation. To be effective, review and feedback need to be carried out regularly.

7.2   Availability of PCM data

The first part of the review process asks the questions:

- Was the required PCM plan carried out?
- Were the data required for PCM assessment delivered?

If the PCM plan/programme was not carried out or if sufficient PCM data were not generated, then a review should be carried out as to why and where the PCM plan failed and what course of action is required to recover the plan and data arising therefrom. It is recommended that this review is formally documented and that a corrective action plan is put in place.

7.3   Data validation

A preliminary assessment of data quality should be made. Should the data be considered corrupted or deficient to the extent that a PCM assessment cannot be performed, then a traceable corrective action plan should be put in place to rectify either the process or equipment responsible for returning degraded or corrupted information.

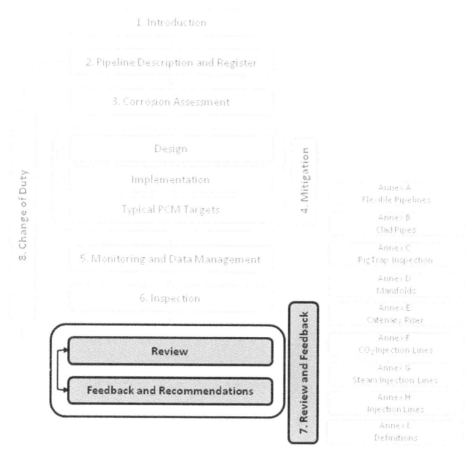

7.1 Position of Chapter 7 within the structure of this RP

7.4   Recommended data

The next part of the review process examines and assesses the data generated. Typical PCM data sets used for this review include, but are not limited to:

- data from corrosion monitoring devices
- data from inspection
- data from sampling
- reliability of chemical injection
- injection water quality
- availability of mitigation systems (cathodic protection (CP), physical treatments)
- historic data (previous status, repairs, etc.)
- modifications
- operational data
- results from corrosion assessments
- other relevant PCM data (3rd party entrants).

7.5   Reporting

The PCM data are used to determine whether key PCM targets are being met. Typical key targets include, but are not limited to:

- Is the corrosion rate within the acceptable range set for the pipeline?
- Is the efficiency of mitigation systems being met, i.e. corrosion inhibitor (CI) injection?
- Are corrosion probes functioning?
- Are corrosion probe readings within the acceptable range set for the pipeline?
- Were corrosion coupons retrieved per the PCM plan?
- Are the data returned by the corrosion coupons within the acceptable range set for the pipeline?

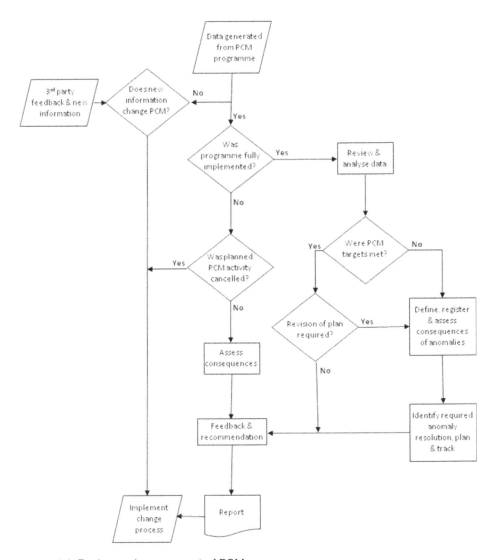

7.2 Review and assessment of PCM

- Was the pipeline subject to in-line inspection (ILI) per the PCM plan?
- Are the data returned from ILI within the acceptable range set for the pipeline?
- Are data arising from any other pipeline inspection activities in line with pipeline PCM targets?
- Are data arising from any pipeline operational pigging activities in line with pipeline PCM targets?

These data are used to compile a report on pipeline condition. The report should typically include:

- Any failure to achieve PCM target anomalies identified
- A statement and plan as to how the anomalies identified are to be rectified
- A schedule for anomaly rectification
- A statement as to whether key PCM targets were met
- A statement as to the adequacy of the PCM scheme for the immediate future
- Recommendations for improvement or changes to the PCM scheme in place.

This report is typically produced once per year for each pipeline system.

8

Change of duty and implementation of change

The function of a pipeline may change at any time throughout its lifetime to allow transportation of alternative fluids. Any change to the fluid conveyed will need a reassessment of the pipeline design to ensure that it is capable of conveying the fluid safely.

This chapter (Fig. 8.1) outlines the necessary steps and information required to implement change of duty in a pipeline system.

8.1 Introduction

Change of inventory and/or operating conditions may occur for the pipeline systems over their design lifetimes. These changes typically arise from [1]:

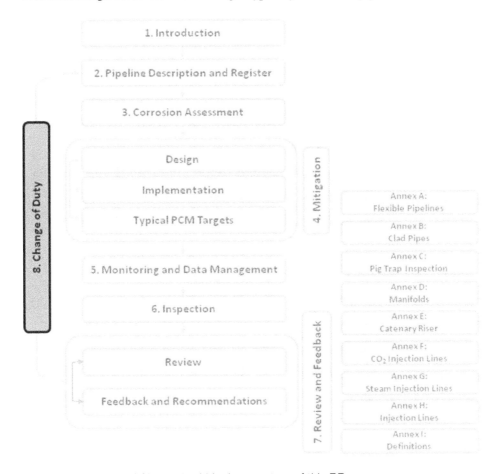

8.1 Position of Chapter8 within the structure of this RP

- Major modifications/remedial work to the pipeline.
- Changes in fluid composition or type: pipelines may be designed to operate with dry gas but changes to the status of offshore installations may only be achieved if the gas can be transported in a wet state – this may have a significant effect on the integrity of those pipelines and downstream facilities.
- Entry of a new product into an existing pipeline system.
- Changes in production rates.
- Changes in safe operating limits, e.g. when changing from one pressure to another.
- End of use of a pipeline: this notification should set out the steps to be taken to decommission, dismantle or 'abandon' a pipeline. It is envisaged that a notification will comprise a timetable indicating when the pipeline is to be taken out of service, how long the line was to remain decommissioned and a description of how the line is to be made permanently safe.
- Changes in pipeline materials and equipment: this may comprise no more than a map or chart showing where the changes are to take place and a brief description of the material and/or dimensional changes.
- Re-routing of pipelines, e.g. in close proximity to offshore installations which could have an effect on the safety of the installation.
- Re-routing of pipeline risers on offshore installations which may then pass closer to living quarters or other vulnerable areas.
- The repositioning of emergency shut down valves (ESDVs) on pipeline risers.
- Change of application (e.g. from injector to producer, etc.).

The above is defined as 'Change of Duty'.

8.2   Management of change

A review should be made to determine how the Change of Duty will affect the originally established corrosion management programme. This will normally follow in-house management of change procedures. However, in the absence of an in-house procedure, aspects that should be taken into consideration in this review can include:

i.   existing pipeline conditions
ii. review of feedback report (Chapter 7)
iii. new operating/duty conditions.

This review should examine all relevant aspects of corrosion management which may include:

- corrosion modelling of the new duty conditions
- compatibility of existing pipeline chemicals with any new chemical used in the original system
- adequacy of existing corrosion mitigating systems
- adequacy of existing pipeline corrosion monitoring facilities for the new operating conditions
- adequacy of the existing pipeline inspection plan
- adequacy of any existing pipeline management schemes.

The results of the Change of Duty review should be formally reported.

The conclusions and recommendations arising from the Change of Duty review should be used to update and revise the pipeline corrosion management (PCM) programme and targets.

Reference

1. UK Health and Safety Executive, 'A guide to the pipelines safety regulations. Guidance on regulations', 1996.

Flexible pipelines are of particular importance in transportation of hydrocarbon from deep water or remote areas. In contrast to rigid pipelines which are essentially simple structures defined by diameter and wall thickness, a typical flexible pipeline may have many concentric layers of plastic or spirally wound steel fibres, each performing a different function. This annex (Fig. A.1) covers and recommends specific corrosion management aspects of flexible pipeline systems within pipeline corrosion management (PCM). It is structured in line with the entire RP and, where appropriate, specifies additional recommendations.

A.1 Introduction

In this flexible pipeline annex, only corrosion of components manufactured from metallic material is considered. Corrosion and degradation of polymer and or composite components/parts are therefore not included.

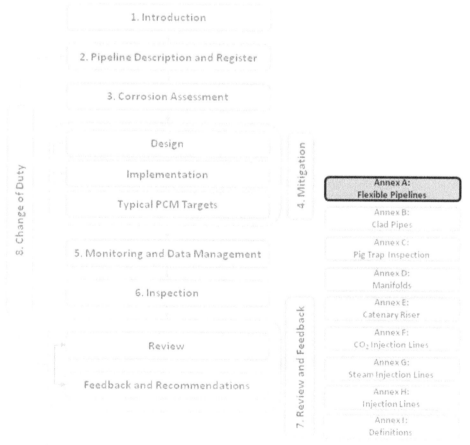

A.1 Position of Annex A within the structure of this RP

Furthermore, in the case of flexible pipelines, flexible pipeline corrosion management (FPCM) is a part of the more global pipe integrity management (PIM) [1] and riser integrity management (RIM) processes and not covered in this annex.

It is recommended that the flexible pipeline manufacturer is consulted and involved in the different steps of the FPCM process. A schematic drawing of a typical flexible pipeline and its components is shown in Fig. A.2.

A.2   Pipeline register and description

Flexible pipeline data on the pipeline description and register (pipe references, environment and pipe cross-section design data) should be as follows:

A.2.1   Basic pipeline register

See general recommendations in this RP.

In addition, inputs required for flexible pipeline design are listed in Section 5 of API 17J.

The following points should be considered:

- Pipeline configuration – in the case of risers (lazy S, free hanging, use of bend stiffener or bellmouth, tethers, clamps and anchors, etc.).
- Conditions at the area of pipe touch-down on the sea-bed (potential for outer sheath wear/damage).
- Name of supplier is an essential requirement.

A.2.2   Physical characteristic register

- Design report reference in the case of existing design
  ○ Structure, material grades, layer thickness and shape
- End fittings configuration

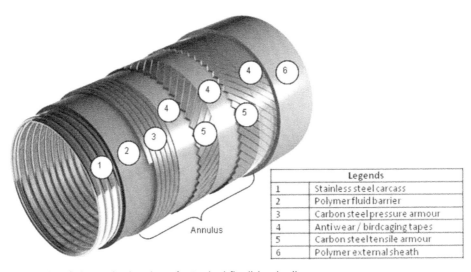

Annulus

Legends	
1	Stainless steel carcass
2	Polymer fluid barrier
3	Carbon steel pressure armour
4	Antiwear / birdcaging tapes
5	Carbon steel tensile armour
6	Polymer external sheath

A.2 Schematic drawing of a typical flexible pipeline

- Ancillaries (bend restrictors (metallic vertebrae), bend stiffeners, buoyancies, tether clamps, etc.)
- Manufacturing data-book reference (data book itself should be accessible by the operating company if needed).

Installation report reference should be noted so that any issues/damage/repairs carried out during installation are clearly identified.

A.2.3 Operating conditions

The key elements of this RP apply equally to this section. In addition, the following parameters should be noted:

- Duration of any storage of the line with treated or untreated seawater should be taken into consideration.
- Annulus environment (based on permeation analysis, and gas vent data, if available (see Assessment Section A.3.1).
- Dynamic service data, if any (including excursions from normal operating conditions).
- Historical data: use for different fields/duties, corresponding durations, storage period, etc.

It should be noted that in the case of reassessment, the operating conditions should ideally be the actual monitoring/inspection data. Otherwise data may be obtained from initial design assessments. The corrosion review process may also lead to a re-assessment of operating data taking account of design rules evolution, lessons learnt from expertise, new know-how and experience development.

A.2.4 Inspection policy and mitigation data

The existing corrosion mitigating measure includes:

- Flexible pipeline structure and materials selection – steel and polymer materials, pipeline configuration and ancillary equipment.
- Inhibition/biocide treatment and other chemicals during seawater storage and during operation (via inhibitor injection in the annulus and/or in the bore).
- Cathodic protection (CP) system: type and design.
- Others, depending on proprietary design.

A.3   Corrosion assessment

In addition to the parameters outlined in Chapter 3, careful consideration should be given to events that might result in (i) breach of the external sheath, (ii) H_2S break-through (as a result of reservoir souring) and (iii) reduced corrosion fatigue life.

It is worth reiterating that corrosion assessment of flexible risers becomes even more important at the design stage and active dialogue between the operating company and the manufacturer is paramount.

A.3.1 Relevant corrosion threats

In flexible pipelines, two distinct types of corrosion threat need to be considered:

- Corrosion threat to the steel carcass which is continually exposed to the bore fluid and generally made of corrosion resistant alloy (CRA).
- Corrosion threats to the reinforcement wires in the annulus which are exposed to the confined annulus environment with no direct contact with the bore fluid [2–6].

The relevant corresponding corrosion threats are given in the Bibliography (Section A.9 and API 17J and 17B).

A.3.2   Gathering data inputs

In the case of flexible pipelines, the required inputs include both (i) the bore environment (in contact with the carcass material and inner part of the end fittings) and (ii) the confined annulus environment (which is in contact with the steel armour materials). The bore environment input parameters and the structure of the flexible pipeline are essential to calculate the annulus conditions (see Sections A.3.2.1, A.3.2.3 and A.3.2.5).

In addition to the basic pipeline description data listed in Section A.2, the following data are necessary to carry out corrosion assessment of flexible pipelines:

A.3.2.1   Fluid chemistry

Corrosivity assessment in the bore environment is similar to the rigid pipelines as outlined in Chapter 3. However, corrosivity assessment in the annulus environment from the bore operating conditions constitutes specific understanding of corrosion phenomenon for flexible pipelines.

Annulus environmental parameters include:

- temperature profile in the annulus
- acid gases (CO_2 and H_2S) partial pressures
- presence of condensed water (yes/no)?
- annulus pH
- seawater flooded annulus by design or following external sheath accidental damage.

A.3.2.2   Gas composition

The key elements of this RP apply equally to this section and those of Section A.3.2.1.

A.3.2.3   Flow regime

The key elements of this RP apply equally to this section.

A.3.2.4   External assessment

The key elements of this RP apply equally to this section.

In addition, stress calculations from the operating conditions will allow prediction of corrosion fatigue performance (Section A.2).

A.3.2.5   Existing third party activities and infrastructure

The key elements of this RP apply equally to this section. However, such data should be gathered and used for both the bore and the annulus conditions.

In addition, incident reports for indication of impact, and outer sheath damage if any should be reviewed.

A.3.3   Competency level required for corrosion assessment

See general recommendations in this RP. Special competencies in the design, manufacture and remediation of flexible pipelines are recommended.

A.3.4   Non-routine operations

Typical non-routine operations, which may be highly critical in the specific case of flexible pipelines (due to thin wall CRA carcass), include acid flushing, hydrotest, factory acceptance test (FAT, if performed with untreated seawater/brackish water), storage of the line in corrosive environment (oxidation in air, seawater storage, etc.). In such events, additional precautionary measures should be considered. Further guidelines with regard to the CRA carcass can be found in Annex B.

A.3.5   Other hotspots

During corrosion assessment, particular attention should be paid to the hotspots described in Table A.1.

If any hotspots appear following an accidental event, which were not taken into account in design, assessment should be carried out immediately. This highlights the importance of regular inspections, monitoring and corrosion review of existing lines as outlined in Chapters 5, 6 and 7.

A.3.6   System corrosivity

Severity/magnitude of the relevant corrosion threats should be evaluated by tests, models, or operating experience or a combination of these three approaches. Further guidance is given in the Bibliography (Section A.9 and API 17J and API 17B).

Table A.1   Potential hotspots in flexible pipelines

Hotspot description	Affected item
Locations where O_2 + water + no/inadequate cathodic protection (CP) may be found: external sheath breach in or close to the splash zone, in splash-zone and above and within the guide tube	Wires – at air/water interfaces
Locations where cyclic dynamic loads apply and water is present (corrosion fatigue critical areas), e.g. topside, touchdown point (TDP)	Wires
Zones where there is a threat of external sheath damage by impact (fishing/vessel interaction, etc.)	Wires + carcass

Table A.2 Potential threats and mitigation measures for flexible pipelines

Mitigation method	Mitigation measure	For the mitigation of:
Materials design	Selection of steel wire	Hydrogen induced cracking (HIC)/ sulphide stress cracking (SSC), corrosion fatigue issues on wires
	Selection steel carcass	Corrosion issues during service or storage
	Pipe structure selection: pressure armour, wire thickness, number of layers, end fittings design, sacrificial sheath, and double sheathes	• Corrosion fatigue (CF) issues • Uniform corrosion issues • HIC/SSC issues
	Pipeline equipment design: stiffener	CF issues by mitigating dynamic loads
	Pipeline configuration selection: free hanging, Lazy S, pliant wave, use of buoyancy modules, tether clamps, etc	CF issues by mitigating dynamic and static loads
CP design	External corrosion	• Accidental O_2 corrosion of steel wires in case of breach in external sheath • Subsea end fittings corrosion
Corrosion inhibition	Bore corrosion inhibition	Bore inhibition (biocide, oxygen scavenger, corrosion inhibitors) may be used in case of flexible pipeline storage in seawater if the storage duration exceeds the duration acceptable for the selected carcass material. In some specific cases, nitrogen flushing of the bore, or rinsing with fresh water may also be recommended in the installation/ operation manual provided by the manufacturer
	Annulus corrosion inhibition	This could include inhibitor injection in the annulus, N_2 flushing of the annulus
	Maintenance/repair of external sheath	Repair of external sheath is an option to avoid the reintroduction of oxygenated water in the annulus. A common way to repair consists of applying a clamp around the external sheath, sometimes injecting a resin or sealer inside the clamp. External sheath patch welding is also used
Flushing	Guide tube flushing	Pressurisation of guide tube is an option to displace water from around the pipeline within the tube if it is determined that there is an external sheath breach at this location

A.4   Corrosion mitigation

The key elements of this RP apply equally to this section. In addition, potential threats and respective mitigation measures are included in Table A.2.

A.5   Monitoring and data management

The key elements of this RP apply equally to this section.

A.5.1   Corrosion monitoring systems for flexible pipelines

A.5.1.1   Corrosion monitoring of the bore environment

No direct corrosion monitoring methods (coupons, corrosion probes, electrochemical methods, etc.) are currently used in the bore of flexible pipelines. There are indirect means of assessing corrosion in the bore. However, these are not in common use and are not covered in this document.

A.5.1.2   Corrosion monitoring within the annulus

Methods of monitoring within the annulus are covered in Table A.3.

Table A.3   Methods of monitoring within the annulus in flexible pipelines

Parameter to be monitored/detected	Method of monitoring
Temperature	Annulus temperature can be monitored using optical fibres
Water flooding detection	Presence of water in the annulus could be monitored by several means such as: • optical fibre systems • annulus vacuum test • annulus nitrogen pressure test (indication of free volume in the annulus) • automatic annulus testing system • hydrogen monitoring of vent gas (see also below)
Annulus gas sampling	The annulus composition may be monitored by annulus gas sampling and analysis; Such analysis may be an indicator of corrosion (presence of H_2) or souring in the annulus environment
Annulus gas flowrate	Annulus gas venting flowrate changes may be used as an indicator of changes in operating conditions or of possible flooding of the annulus
CP potential	CP potential gives information on anode consumption (it should be noted that several other corrosion monitoring devices may be in development or not fully qualified but available)
Corrosion rate monitoring	No method of direct corrosion rate monitoring (coupons, corrosion probes, electrochemical methods, etc.) is currently used in the annulus of flexible pipelines
Stress/strain monitoring	This is useful in reassessing the remaining corrosion fatigue life. Furthermore, this may provide an indication of corrosion damage of reinforcements. Such systems are proprietary and require discussion with providers for their use

A.5.2   External corrosion monitoring

The key elements of this RP apply equally to this section. In addition, those outlined in Section A.5.1 should be considered.

A.5.3   Data management

The key elements of this RP apply equally to this section.

A.6   Inspection

A.6.1   Inspection tools and purpose

The main inspection method is visual surveys of the external sheath. The most practical means of assessing the annulus free volume is by vacuum test. Pressure testing of the annulus is not normally recommended as it may lead to bursting of the external sheath, unless carried out with extreme caution. Direct assessment of the status of the reinforcement in the annulus (such as riser scanning, wire extraction and/or inspection, etc.) and the carcass is particularly challenging and will not be dealt with in this RP.

A.6.2   Inspection after line recovery: specific case of corrosion expertise

Corrosion understanding is sometimes achieved after the recovery of flexible pipe-lines. Such expertise provides valuable feedback from the field which should thus be used in the PCM process. In some instances, flexible pipelines can be recovered and reused – in such cases, full inspection and reassessment considering the history of the pipeline and the new design inputs is necessary before redeployment. At the end of life, destructive examinations can provide valuable information to assist in improving design and construction.

A.6.3   Baseline inspection

The key elements of this RP apply equally to this section.

A.6.4   Use of risk-based inspection (RBI)

The key elements of this RP apply equally to this section.

A.6.5   Data management

The key elements of this RP apply equally to this section.

A.7   Review

The key elements of this RP apply equally to this section.

A.8   Change of duty

The key elements of this RP apply equally to this section.

A.9   Bibliography

- API 17J/ISO 13628-2, Specification for Unbonded Flexible Pipe.
- API 17B/ISO 13628-11, Recommended Practice for Flexible Pipe.

A.10   References

1. E. Binet, P. Tuset and S. Mjoen: Proc. Offshore Technology Conference 2003 (OTC 2003), 5–8 May 2003, Houston, TX, Paper 15163.
2. F. Ropital, C. Taravel-Condat, Jn. Saas and C. Duret: Proc. EUROCORR 2000, 10–14 September 2000, London, UK.
3. C. Taravel-Condat, M. Guichard and J. Martin: Proc. Offshore Mechanics and Artic Engineering Conference, 8–13 June 2003, Cancun, Mexico.
4. A. Rubin and J. Gudme: CORROSION 2006, NACE Annual Corrosion Conference, 12–16 March 2006, San Diego, CA, Paper 06149.
5. R. Clements: Proc. EUROCORR 2008, 7–11 September, Edinburgh, UK.
6. E. Remita, F. Ropital and J. Kittel: CORROSION 2008, NACE Annual Corrosion Conference, 16–20 March 2008, New Orleans, LA, Paper 08538.

This annex (Fig. B.1) covers and recommends specific corrosion management aspects of solid corrosion resistant alloy (CRA) and CRA clad or lined and non-metallic lined pipeline systems within pipeline corrosion management (PCM). It is structured in line with the entire RP and, where appropriate, specifies additional recommendations. For the backing materials, where exposed to an environment, the main recommendations of this RP should be followed.

B.1 Introduction

Solid CRA and CRA clad or lined and non-metallic lined pipelines are being used in hydrocarbon production and water injection systems for handling corrosive media.

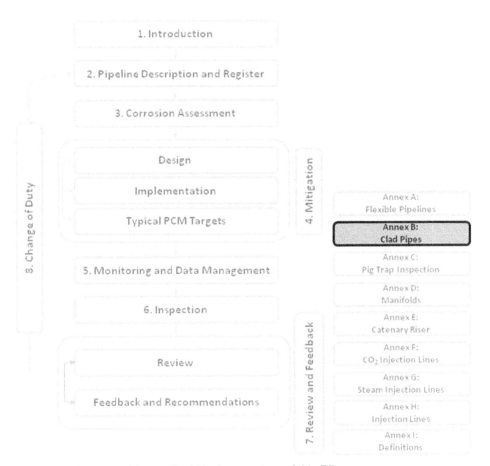

B.1 Position of Annex B within the structure of this RP

They provide a number of corrosion mitigating measures in response to increasing system corrosivity. These options are typically selected if system corrosivity cannot be addressed adequately by other means. Clad or lined pipelines are formed from a C-steel pipe lined internally or externally in some cases, with a layer of CRA or non-metallic material. The clad or liner is not normally considered to act as a pressure retaining item.

B.1.1   Description

The typical process of manufacturing clad/lined pipes includes:

i.   Metallurgically bonded clad lines: CRA internally clad linepipe is typically pro-duced using CRA clad onto the C-steel plate. The clad plate is most commonly manufactured by roll bonding the CRA cladding onto the base plate (backing steel), although it may also be produced by explosion bonding. In each case, a metallurgical bond is formed between the C-steel and the CRA layer. Alternative processes include CRA weld overlay onto a C-steel pipe. Clad CRA layers are typically 3 mm thick.

ii. Mechanically bonded clad lines: CRA internally lined linepipe commonly utilises a CRA liner that is hydraulically expanded onto a C-steel outer pipe and seal welded at the ends. In some instances, the end sections of the lined pipe are weld clad, principally to provide for easier inspection of girth welds during laying. For lined pipe, the majority of the pipe length relies on a mechanical bond being formed between the C-steel and the CRA layer. Clad CRA liners are typically thinner than the clad option.

iii. Non metallic lined lines: a non-metallic liner is inserted into a C-steel pipe and provides a barrier between the internal environment and the steel. Non-metallic liners are typically >10 mm thick and have a tight fit to the backing material.

B.1.2   Examples

Examples of commonly used CRAs for flowlines and pipelines, together with their typical threats, are given in Table B.1. Table B.2 summarises the status of non-metallic liner technology. Both of these tables offer general guidance. However, for each specific application, a materials performance evaluation should be conducted to ensure suitability for the intended service.

B.2   Pipeline description and register

For CRA clad/lined pipes, in addition to the data for the backing steel, inventory data should include the type of clad/lined/solid pipe, material type and chemistry of the cladding/liner and its respective thickness.

   The non-metallic lined pipe register should include details of the backing steel and the liner material type and method of installation and joining.

B.3   Corrosion assessment

B.3.1   Relevant corrosion threats

Whilst general corrosion is not typically a significant threat for CRA linepipe, localised corrosion can be an issue. Similarly, if a field sours in later life then

Table B.1 Commonly used CRAs in flowlines and pipelines

Generic type	Examples	Typical composition (wt.%)				Typical practical limitations on use	Typical usage
		Cr	Ni	Mo	Others		
Super martensitic stainless steel	Super 13%Cr (weldable)	11.5–13.5	4.5–6.5	2–3		• Some susceptibility to chloride stress corrosion cracking (SCC) at elevated temperatures • Pitting and crevice corrosion in chloride-containing waters with residual oxygen • SSC under some conditions [1] • Hydrogen cracking under some conditions	Used as solid CRA only
Austenitic stainless steel	316L	18	12	2.5		• Chloride SCC threat in the presence of oxygen and/or H$_2$S at elevated temperatures • Pitting corrosion at lower temperatures	Used as liner, cladding, overlay or solid
Duplex stainless steel	22%Cr Duplex	22	5.5	3	N	• SCC risk at higher temperatures • SSC under some conditions • Hydrogen cracking under some conditions • Pitting and crevice corrosion in chloride-containing waters with oxygen	Used as liner, cladding, overlay or solid
	25%Cr Superduplex	25	7	3.5	N, Cu, W	• SCC risk at higher temperatures • SSC under some conditions • Hydrogen cracking under some conditions • Pitting and crevice corrosion in chloride-containing waters with oxygen at higher temperatures	Used as liner, cladding, overlay or solid
Nickel base alloy	Alloy 625	22	60	9	Fe, Nb	Crevice corrosion under some conditions	Used as liner, cladding or overlay

Table B.2 Commonly used non-metallic liners and their typical application and limits

Application	Options	Current status	Typical P, T, limits
Onshore – WI (New build & Rehab)	Spoolable slip liner	Field proven	2500 psi PE < 60°C, PEX < 90°C
Onshore – Production	Plain tight fit liner	• Field proven for hydrocarbon liquids	Gas: P < 200 psi No gas: PE < 50°C; PA-11 <90°C
(New build & Rehab)	Grooved tight fit liner	Field proven	PE < 60°C, 3000 psi PA-11 < 90°C, 6000 psi
	Spoolable slip liner	Field proven	2500 psi PE < 50°C; PA-11 < 90°C
Offshore – WI (New build)	Plain tight fit liner	• Field proven	PE < 60°C
Offshore – Production (New build)	Plain tight fit liner	Field proven for hydrocarbon liquids	Gas: P < 200 psi No gas: PE < 50°C; PA-11 < 90°C
	Grooved tight fit liner	Reeled only	PE <60°C, 3000 psi PA-11 < 90°C, 6000 psi
Offshore – WI (Rehab)	Plain tight fit liner	Concept proven	PE < 60°C
	Spoolable slip liner	Field proven	2500 psi PE < 60°C, PEX < 90°C
Offshore – Production (Rehab)	Plain tight fit liner	Concept proven for hydrocarbon liquids	Gas: P < 200 psi No gas: PE < 50°C; PA-11 <90°C
	Grooved tight fit liner	Concept proven	PE <60°C, 3000 psi PA-11 <90°C, 6000 psi
	Spoolable slip liner	Field proven	2500 psi PE < 50°C; PA-11 <90°C

Notes:
1. E = polyethylene; PEX = cross linked polyethylene; PA = poly amide; WI = water injection.
2. Legends:
   • Technologies (unless underlined) are proven, to the limits stated.
   • Technologies underlined indicate further development or trials required.
3. High water cut production may on detailed analysis have the same limits as 'WI' in this chart, if associated gas partial pressures are low.
4. Perforated liners proven in JIP flow loop testing, but technology on hold due to corrosion concerns.

operating companies should ensure that the pipeline cladding/lining is suitable for sour service, following the guidance in NACE MR0175/ISO 15156-3 for the avoidance of sulphide stress cracking (SSC)/stress corrosion cracking (SCC). The threat of SCC is increased for 13%Cr family and 316L alloys if oxygen ingress occurs, necessitating careful oxygen control of any injected chemicals.

SSC and hydrogen induced cracking (HIC) qualification of the backing steel of metallic clad/lined pipe for sour service application is not normally carried out.

However, this may be a consideration for some applications. Gas permeation is a consideration for non-metallic liners and therefore, SSC/HIC qualification of the backing steel may be required.

Any connections between CRA and non-CRA metallic equipment should be considered with respect to the threat of galvanic corrosion.

Typical solid CRA submarine pipeline materials can be susceptible to hydrogen embrittlement as a consequence of external cathodic protection (CP). This should be considered in the design and operation of such pipelines.

For CRA clad, API 5LD should be followed with respect to corrosion assessment of the CRA layer.

The degradation threats for non-metallic liners include, but are not limited to:

- chemical attack and adsorption
- gas permeation
- temperature and pressure limits
- ageing
- blistering of the liner due to gas permeation.

In addition, gas can permeate through the liner, building up at the interface with the steel backing during operation. Depressurisation of the pipeline results in expansion of the gas at the interface and collapse of the liner.

B.3.2   Gathering data inputs

The key elements of this RP apply equally to this section. In addition, for non-metallic liners, temperature and pressure limits of the plastic liner should be made available.

B.3.3   Competency level requirements

The key elements of this RP apply equally to this section.

B.3.4   Non-routine operations

CRA linepipes can be susceptible to localised corrosion in several incidental environments (especially lower grades).

Similar precautions as in the main body of this document should be taken to minimise the risk of corrosion from unspent acids for solid and clad/lined CRA and non-metallic lined pipes.

Common CRA pipelines (including solid and clad/lined options) are particularly susceptible to localised oxygen corrosion in hydrotest water or in the case of seawater ingress.

Some CRAs may be subject to much more severe localised O_2 induced corrosion than C-steels. Therefore, the water quality and treatment (source, chemical composition, oxygen content, added chemicals, etc.) used for hydrotesting, and quick dewatering and drying, may become more critical. Considerations should be given to biocide treatment of the water to avoid microbial induced corrosion (MIC). Precautions should also be taken to avoid raw seawater ingress during installation and tie-in operations. If raw seawater enters the pipe system, it should be removed by flushing the system with fresh water within an appropriate period to prevent corrosion damage.

B.3.5 Other hotspots

Notable hotspots for CRA options are as follows:

- Particular attention should be given to avoid galvanic corrosion when connecting CRA sections to C-steel.
- Externally localised corrosion is a particular threat in splash and atmospheric zones for solid CRA lines in which cases appropriate coating systems should be considered.

B.3.6 System corrosivity

The external corrosion threat assessment of clad/lined pipe should consider the C-steel backing material (and seam/girth welds) only. Thus, the external corrosion threat assessment should be largely similar to that for a C-steel pipeline as covered in this RP. The internal corrosion threat assessment should consider the potential for localised corrosion (including under-deposit corrosion) and environmentally-assisted cracking of the CRA. Similarly, the threat posed by gas permeation through the non-metallic liner should be considered.

Whilst internal general corrosion may not be a significant threat, the potential for erosion should be considered. This is especially important for clad and lined options since, after loss of the corrosion resistant layer, rapid corrosion of the backing C-steel can occur. Potential synergy between erosion and corrosion should be considered and this is particularly important for 13Cr martensitic/supermartensitic linepipe, which may not be capable of maintaining a passive surface in erosive conditions. Hydrogen embrittlement is a potential threat to 13Cr martensitic/supermartensitic and duplex/superduplex stainless steel submarine pipelines under CP. The magnitude of the threat depends upon the following factors:

- stress
- microstructure
- applied CP
- coating quality.

Doubler plates fillet welded to linepipe for direct anode attachment are particularly problematic since these provide high residual stress directly at the site of the applied CP and necessitate a (often poor quality) field coating.

B.4 Corrosion mitigation

There is generally no need for mitigation against internal general corrosion, since solid CRA, CRA cladding/lining or non-metallic lined pipe can be assumed to be immune. However, actions may still be required to minimise the build-up of solids, prevent formation of free sulphur, etc. Mitigating actions for external corrosion of clad/lined pipe are the same as for C-steel pipelines. For solid supermartensitics, duplex or super duplex submarine lines, additional precautions are required for minimising the risk of hydrogen embrittlement. This can be mitigated by following the guidelines provided in the Bibliography (Section B.9 and EEMUA 194 and DNV RP F112 for duplex and super duplex pipelines).

Typically, non-metallic liner collapse due to decompression is mitigated for hydrocarbon lines by the manufacture of grooves in the polymer and bleeding off the permeated gas periodically through nipples in the steel backing.

B.4.1   Materials selection

Typical clad/liner options include those outlined in Table B.1. Selection should be based on compatibility with internal as well as external environments. Thus, specific testing is required to qualify pitting, crevice, SSC and SCC performance. It should be noted that both claddings and linings may demonstrate differing corrosion performance compared with a conventional solid CRA, since they have very different processing routes. Consequently, it should be verified that the manufacturing route does not adversely affect the corrosion performance of the CRA layer. For example, welding and post-weld heat treatment (PWHT) require a high level of qualification and QA to avoid embrittlement and/or degradation of corrosion-resistance due to the formation of detrimental microstructures.

Materials selection for non-metallic liners is made according to the operating and design conditions, and the required mechanical and chemical resistance properties to achieve functionality throughout the operational life.

The main materials selection criteria for water injection (WI) and production (multiphase hydrocarbon fluid or gas) service include, but are not limited to:

- chemical, mechanical and thermal resistance to operating and external environment
- expected service life
- compatibility with installation method and joining techniques
- materials processability
- economics.

The most commonly used liner material is high-density polyethylene (HDPE), whilst cross-linked polyethylene (PEX) and polyamide 11 (PA-11) are seldom used.

HDPE is the most commonly used material for water injection flowlines, and multiphase oil and gas gathering and flowlines. There are a number of grades, with small variation in properties, all specified as outlined in the Bibliography (Section B.9). PE options are typically capable of ca.65°C in water and ca.50°C in hydrocarbon.

PEX has improved mechanical (tensile and creep), thermal (operating temperature) and similar chemical resistance properties compared with HDPE. The disadvantage is the difficulty of fusion welding PEX. Temperatures of up to 90°C in water can be acceptable. Not commonly used.

PA-11 (low plasticiser and methanol washed) is a high performance material primarily designed, and used, for multiphase oil and gas gathering and flowlines. Due to issues with plasticiser leaching out and attack by methanol, a low plasticiser and methanol washed grade was developed. Temperatures of up to 90°C in hydrocarbon can be acceptable, although not commonly used.

B.4.2   Internal and external coatings

Lower grade CRAs may require internal protection during storage/transportation. External coating requirements for clad/lined pipe are identical to those of C-steel pipelines. External coatings should also be considered at hotspots and for solid CRA submarine linepipes to reduce current drain on the CP system and to mitigate the threat of hydrogen embrittlement.

B.4.3   Chemical and physical mitigation methods

The potential consequences of oxygen ingress in CRA pipelines are typically greater than those for C-steel pipelines, since rapid localised corrosion and/or SCC can occur. Therefore, care is required to ensure that any injected chemicals are managed to avoid oxygen contamination.

In the particular case of water injection, materials selection and water quality need careful consideration in respect to residual O_2 and chlorine to avoid localised corrosion of CRAs. Further guidance can be found in Annex H.

A corrosion allowance is not typically required for solid CRA, clad or lined pipelines. Where sand/solids production is significant, a wastage allowance may be required. For weld overlay lines, it should be ensured that adequate clad thickness (with the appropriate chemistry) is present throughout the life.

It should be noted that in certain circumstances (onshore buried lines), corrosion allowance may be required to cater for the potential external corrosion threat.

B.4.4   Operational pigging

Operational pigging may still be required for flow assurance. In such a case, care should be taken to choose appropriate pigs to avoid damage to the CRA layer or non-metallic liner. Even when appropriate pigs are selected, the potential for damage caused by entrained solids (accumulated during operation) swept ahead of the pigs should be assessed.

B.4.5   Competency

The key elements of this RP apply equally to this section.

B.5   Corrosion monitoring and data management

The key elements of this RP apply equally to this section.

B.6   Inspection

The key elements of this RP apply equally to this section.

In addition, for solid CRA, clad or lined lines, inspection for internal corrosion is normally not required. However, specialised inspection may be required where erosion is a concern and where conditions are identified for localised corrosion/damage.

B.7   Review

The key elements of this RP apply equally to this section.

B.8   Change of duty

The key elements of this RP apply equally to this section.

B.9   Bibliography

- API Spec 5LD, Specification for CRA Clad or Lined Steel Pipe.
- ISO 3183, Petroleum and Natural Gas Industries – Steel Pipe for Pipeline Transportation Systems.

- API 5LC, Specification for CRA Line Pipe.
- ASTM D2239, Standard Specification for Polyethylene Plastic Pipe Based on Controlled Inside Diameter.

B.10   Reference

1. 'Petroleum and natural gas industries – Materials for use in H_2S containing environments in oil and gas production'. MR0175/ISO 15156, National Association of Corrosion Engineers, Houston, TX.

All aspects of corrosion management of pig launchers and receivers are covered within the pipeline corrosion management (PCM) programme as an integral part of the pipeline management system (PMS). However, specific additional recommendations are made in this Annex (Fig. C.1) in respect of inspection and maintenance only.

C.1   Inspection

- The end closures and valves should be included in the PCM programme and maintained regularly to assure fitness for purpose. Maintenance of subsea systems is challenging and this should be recognised in the design of the subsea facilities.

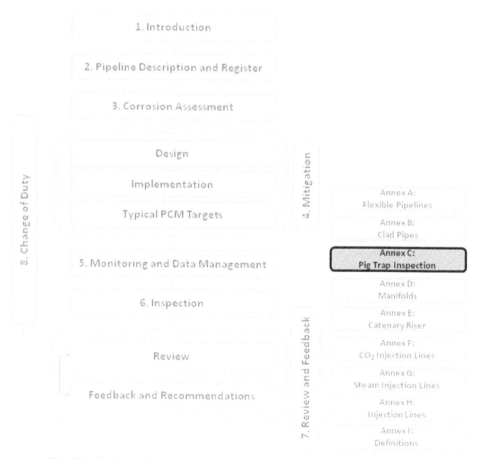

Fig. C.1 Position of Annex C within the structure of this RP

- Particular attention should be given to potential defects in any securing bolts, nuts, and nut housing, their method of attachment, and build-up of corrosion products that interfere with the correct operation of the mechanism.
- Particular attention should be paid to locations where water/moisture can collect as these are prone to crevice corrosion threat.
- Dismantling and close visual inspection of main load bearing components for any sign of corrosion or cracking should be undertaken regularly.

Manifolds are more complex components than the main pipeline and require additional consideration with respect to corrosion management. By their nature, subsea manifold systems require yet more complex and expensive intervention than equivalent offshore or onshore surface facilities hence the pipeline corrosion management (PCM) programme needs to deliver integrity with minimum intervention. For the purposes of this RP, subsea systems will include pipeline end terminations (PLET) and pipeline end manifold (PLEM) arrangements and any associated protective structures. This annex (Fig. D.1) outlines the necessary steps required specifically in relation to manifolds which are over and above the PCM programme outlined within the RP.

D.1 Position of Annex D within the structure of this RP

D.1   Introduction

Manifold systems are designed to:

i.   gather production fluids from different wellheads and transfer into the production pipeline

ii.   distribute injection water or gas to wellheads as part of an enhanced recovery or disposal system.

Manifolds will typically be made up of a complex arrangement of piping valves and instrumentation, and for subsea systems, additional control modules and their associated fittings contained within a protective steel structure. Further details can be found elsewhere outlined in the Bibliography (Section D.9).

The following sections discuss additional or specific requirements for the corrosion management of manifold systems to those addressed within the main body of the RP or in applicable annexes.

D.2   Pipeline description and register

The key elements of this RP apply equally to this section.

D.3   Corrosion assessment

Manifold systems are exposed to both internal and external corrosive environments but due to their layout, effective corrosion management is more difficult to achieve and measure as a result of:

- complex piping and valve arrangements leading to transient flow regimes which impact internal corrosion assessment and effectiveness of chemical treatment
- larger number of hub/flange connections compared to the pipeline giving an increased corrosion threat at interfaces (internal and external) and potential leak paths
- potentially a mix of different materials of construction and a greater concern from galvanic corrosion
- inability to undertake intelligent in-line inspection (compared to the pipeline) as a direct corrosion monitoring method.

Subsea systems present particular challenges including:

- the need to ensure effective cathodic protection (CP) across a variety of interfaces and materials with different polarisation potentials
- risk arising from hydrogen embrittlement of certain corrosion resistant alloys (CRAs)
- the impact of seawater flooding as subsea systems may be exposed for long periods during installation or commissioning causing localised pitting or crevice corrosion of non-resistant components. Valves are particularly susceptible since they typically contain a variety of materials with lower corrosion resistance. Ancillary components, often manufactured from thin wall tubing, such as control panel tubing, chemical injection tubing, instruments, etc., may be particularly vulnerable.

D.4   Mitigation

The key elements detailed in Chapter 4 apply equally to this section.

The use of C-steel for managing corrosion from production fluids will typically rely on chemical treatment (unless the design life of the field is short and/or the fluids considered benign) and limited to onshore facilities or offshore topside facilities where monitoring and inspection activities can be managed effectively.

Manifolds constructed from CRA are often considered as the most robust and cost effective material selection for managing production fluids and injection water/gas, particularly for subsea systems. They do not require chemical treatment or corrosion monitoring (although the likelihood of external stress corrosion cracking (SCC) must be considered in selecting the optimum grade).

The most common CRA materials are either solid duplex stainless steel or C-steel clad/weld overlaid with Alloy 625, although the final selection should also consider material availability, ease of welding and inspection during fabrication, potential galvanic corrosion between different materials, design features to minimise stress discontinuities, suitability under CP, etc. This may also include the selection of duplex stainless steels with an adequate microstructure (austenite spacing).

Additional recommendations for clad pipe can be found in Annex B.

For subsea systems, the integrity of the protective coating system is particularly important not just for external corrosion protection but to limit the uptake of hydrogen (generated as a result of the CP system) and potential for embrittlement of certain CRAs and dissimilar metal welds. As such, the selection of suitable systems and its application should be given specific attention.

Effective CP for subsea manifolds needs electrical continuity across all components in the system, including bolts on flanged joints, and particular attention should be given to ensure electrical continuity testing during factory acceptance test of these manifolds. The different polarisation potentials of the various materials also need to be considered to avoid the likelihood of 'overprotection' of high strength steels and CRAs. As water depths increase, the difficulty of designing and operating an effective CP system also increase and where available, site-specific data should be used to optimise the design – where such data do not exist, a conservative approach to defining the final current requirement demand, relative to that used for shallower water, is recommended.

D.5   Monitoring and data management

The key elements of this RP apply equally to this section.

In addition, due to their more complex arrangement, manifolds are more difficult to monitor than the pipeline and as such, rely more on indirect than direct methods of corrosion monitoring to demonstrate effective corrosion management. Where feasible, direct methods should target areas where corrosion is predicted to be greatest, e.g. changes in flow direction or velocity, stagnant areas, dead legs, etc.

D.6   Inspection

The key elements of this RP apply equally to this section.

Since it is not possible to undertake in line inspection (ILI), inspection is typically limited to the use of targeted UT techniques where corrosion is predicted to be greatest (e.g. changes in flow direction or velocity, stagnant areas, dead legs, etc.) and appropriate CP surveys.

In the case of subsea manifolds, effective inspection may be difficult due to:

- difficulty of access as result of the protective steel structure
- the presence of a significant number of small and fragile components (such as circlips, gaskets, set-screws, etc.).

As such, inspection is typically limited to periodic diver or remotely operated underwater vehicle (ROV) surveys (visual examination for physical damage or coating breakdown, and effectiveness of the CP system).

D.7   Review and feedback

The key elements of this RP apply equally to this section.

D.8   Change of duty

The key elements of this RP apply equally to this section.

D.9   Bibliography

- EEMUA Publication 194, Guidelines for Materials Selection and Corrosion Control for Subsea Oil and Gas Production Equipment.
- DNV RP F112, Design of Duplex Stainless Steel Subsea Equipment Exposed to Cathodic Protection.
- NACE 7L192, Cathodic Protection Design Considerations for Deep Water Projects.

The steel catenary riser (SCR) concept has been used in deepwater field development as an alternative to other forms of rigid or flexible pipes. This annex (Fig. E.1) covers and recommends specific corrosion management aspects of catenary risers within pipeline corrosion management (PCM). It is structured in line with the entire RP and, where appropriate, specifies additional recommendations.

E.1   Introduction

Catenary risers can be employed between subsea equipment and floating production systems. They should have the required flexibility to cope with the loads induced by movements of the floating system relative to the static infrastructure on the seabed

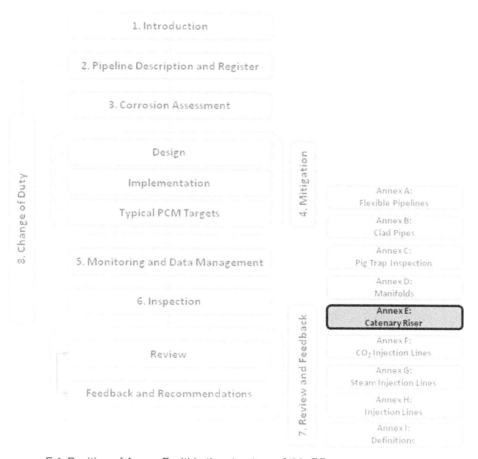

E.1 Position of Annex E within the structure of this RP

whilst retaining full pressure bearing capability. Catenary risers are commonly constructed from C-steel, although corrosion resistant alloys (CRAs) have also been employed successfully. In addition to the riser, there is associated equipment including a flex joint (with a combination of steel and elastomer construction) or stress joint (solid metal construction – either steel or titanium) at the interface with the floating system. Connections between riser sections and end connections to the seabed and surface equipment will also be present, in addition to any required buoyancy modules. Further details can be found elsewhere outlined in the Bibliography (Section E.9).

E.2   Pipeline register and description

The key elements of this RP apply equally to this section.

In addition, the inventory should also include the range of materials used in the construction of catenary risers – these include metals, elastomers and coatings (for corrosion resistance or protection from stray cathodic protection (CP)) used in the connections, the stress joint or flex joint and any associated buoyancy equipment.

E.3   Corrosion assessment

E.3.1   Relevant corrosion threats

Fatigue loading of a catenary riser is clearly a significant driver for design and successful service. However, fatigue and corrosion can act in synergy (corrosion–fatigue) and this will have a dominant influence on the life of the system, potentially greater than any considerations of either fatigue or corrosion in isolation.

It should be noted that corrosion loss and remaining fatigue life may have a complex interaction and should be considered in the design stage.

Where titanium is employed (either for the riser itself or for the stress joint), CP at a level required to protect steel infrastructure will result in hydrogen absorption and embrittlement. Any titanium components must be isolated from the CP system. Furthermore, titanium is susceptible to stress corrosion cracking in anhydrous methanol. Care should be taken to ensure that the water content of any injected methanol is sufficient to inhibit methanol stress corrosion cracking (SCC) of titanium.

The environmental performance of elastomers in flex joints may be a limiting factor for severe service duties. In such instances, stress joints may be preferable.

E.3.2   Gathering data inputs

The key elements of this RP apply equally to this section. In addition, details of cyclic loading (including waves, vortex induced vibrations, floating vessel movements and other environmental parameters) and movements are required for critical positions within the riser system.

The potential for movement in the attached flowlines should be noted.

E.3.3   Competency level requirements

Catenary risers are challenging for design and operation and require specialised skills and competencies.

E.3.4   Non-routine operations

The key elements of this RP apply equally to this section. However, due to the variety of materials that may be present, additional care should be taken when considering non-routine operations to assess the potential degradation of elastomers and galvanic corrosion. Acid compatibility (in particular HF) should be considered when selecting acids which may come into contact with any titanium structures.

E.3.5   Other hotspots

The key elements of this RP apply equally to this section. In addition, the top of a catenary riser is in a splash zone and should be designed accordingly.

E.3.6   System corrosivity

The key elements of this RP apply equally to this section. In addition, there is a requirement for assessment of fatigue and corrosion–fatigue threat. This will require appropriate use of corrosion–fatigue data and more detailed consideration of the influence of inhibitor availability. Localised corrosion threats should be considered as posing a greater risk for catenary risers, since localised corrosion damage can act as the initiation sites for corrosion-fatigue cracks, decreasing the life of the riser. Both the internal and external environments should be considered.

E.4   Corrosion mitigation

The key elements of this RP apply equally to this section. In addition, titanium equipment should be electrically isolated from adjacent components under CP, using insulating joints or flange gaskets and appropriate isolation of the bolts. Appropriate thick external coatings should also be considered to avoid exposure to stray currents.

Also, where titanium stress joints are employed, care must be taken in ensuring CP isolation, to minimise the potential risk of hydrogen embrittlement.

E.4.1   Materials selection

The selection of materials and design not only depends on service conditions, but also on weight implications and the fatigue loading regime. It should be noted that all materials likely to be selected show a corrosion–fatigue synergy, even where corrosion alone would not be expected. Material choices can include C-steel, clad C-steel and CRA for fluids produced and polymer lined C-steel for water injection service.

E.4.2   CRA clad and non-metallic liners

The key elements of this RP apply equally to this section. In addition, thick rubber or similar coatings should be considered for application to titanium stress joints, to minimise the threat of exposure to CP or galvanic coupling from/with connected steel structures.

E.4.3 Chemical and physical mitigation methods

SCRs (without an internal cladding) are likely to rely on corrosion inhibition to achieve an adequate life (including corrosion fatigue). This may place additional demands on the availability of chemical injection programmes.

E.5   Corrosion monitoring and data management

The key elements of this RP apply equally to this section. In addition, monitoring of strain, displacements and vibrations may be required at different positions on the catenary riser and associated equipment. Typical monitoring systems include strain gauges, global positioning systems (GPS), inclinometers and accelerometers, with selection based on the type of data required, the equipment being monitored and the location of concern.

E.6   Inspection

The key elements of this RP apply equally to this section. However, given the criticality of a catenary riser, more frequent inspection may be required than for a pipeline or flowline.

E.7   Review and feedback

The key elements of this RP apply equally to this section.

E.8   Change of duty and implementation of change

The key elements of this RP apply equally to this section.

E.9   Bibliography

Typical design codes

- API RP 2RD, Design of Risers for Floating Production Systems (FPSs) and Tension-Leg Platforms (TLPs).
- API RP 1111, Design, Construction, Operation, and Maintenance of Offshore Hydrocarbon Pipelines (Limit State Design).
- DNV-OS-F201, Dynamic Risers.

With the growing environmental constraints, global warming and public awareness, there is an increasing motivation to reduce carbon emissions. One approach to achieving this is through CO$_2$ capture and storage (CCS). Once captured and compressed, CO$_2$ must be transported to a long-term storage site for which pipelines play a significant role. This annex (Fig. F.1) covers specific corrosion management aspects of CO$_2$ injection lines within the pipeline corrosion management (PCM) plan. It is structured in line with the entire RP and, where appropriate, specifies additional recommendations.

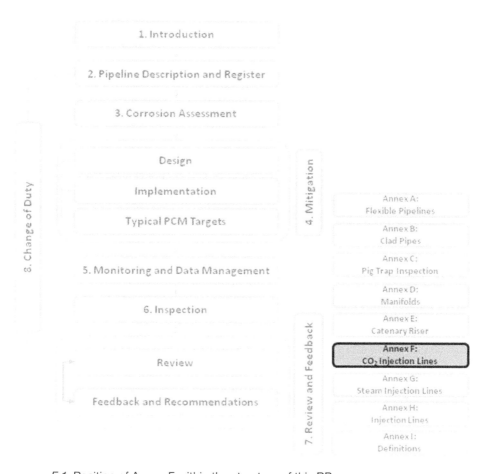

F.1 Position of Annex F within the structure of this RP

F.1   Introduction

It is recognised that, in time, CO_2 pipeline networks, similar to those used for natural gas transmission, will be built, if CCS becomes widely used. Management of such lines is therefore, an important theme and is the subject of the present annex. Further details can be found elsewhere outlined in the Bibliography (Section F.9).

F.2   Pipeline description and register

The key elements of this RP apply equally to this section.

F.3   Corrosion assessment

CO_2 can exist in a variety of physical states depending on the temperature and pressure – this in turn affects system corrosivity. Dense phase CO_2 is not normally corrosive which can allow utilisation of C-steel. However, the picture is complicated and subject to many complementary parameters including the presence of oxidising agents, residual combustion products and H_2S [1].

F.3.1   Relevant corrosion threats

Whilst general corrosion is not typically a significant threat for CO_2 injection lines, localised corrosion can be an issue depending on water content and gas composition.

For CO_2 streams derived from waste gas systems, oxidising species such as NO, NO_2, SO_2, and SO_3, can cause severe corrosion when combined with water, even below the water saturation level.

F.3.2   Gathering data inputs

In addition to the data outlined in Section F.3.1, other parameters that should be compiled include water content and oxidising agents (NO, NO_2, SO_2, SO_3 and O_2).

Furthermore, as CO_2 physical status and the presence of water affect system corrosivity, thermodynamic analysis is recommended to confirm these throughout the pipeline over the operational life.

F.3.3   Competency

The key elements of this RP apply equally to this section.

F.3.4   Non-routine operations

The key elements of this RP apply equally to this section.

F.3.5   Other hotspots

The key elements of this RP apply equally to this section.

F.3.6   System corrosivity

In the absence of free water, corrosion is not an issue in CO_2 injection lines irrespective of CO_2, H_2S or O_2 concentrations. In contrast, in the presence of water, the system will experience severe corrosion. In such a situation, a more prudent approach should be adopted and thermodynamic analysis to ensure lack of water throughout the system and throughout the design life becomes even more essential. In practice, injection of CO$_2$ especially with oxidising agents should not be contemplated unless the absence of free water can be properly maintained.

It should be noted that the corrosion mechanism at high CO$_2$ partial pressure conditions prevailing in CO$_2$ transmission lines, differs from that occurring at lower pressures of CO$_2$. Therefore, conventional CO$_2$ corrosion prediction models (covered in Chapter 3) which relate to the relatively low CO$_2$ partial pressure characteristics of hydrocarbon production conditions are not applicable to CO$_2$ transmission lines. Application-specific testing to determine potential corrosivity should be undertaken.

See also the general recommendations in this RP.

F.4   Corrosion mitigation

Depending on the physical status of CO$_2$, C-steel can be used in the majority of applications for CO$_2$ transmission, particularly when in dense phase and in the absence of water. Therefore, as far as is operationally possible, CO$_2$ physical state should be modified to allow the use of C-steel. This mitigation method to allow the use of C-steel includes:

- total dehydration
- transporting CO$_2$ in 'dense phase', i.e. liquid or supercritical state (>80 bar but normally ~110 bar).

Contaminant levels differ according to the source of CO$_2$ and its processing. Hence, water solubility (and its limits) should be established on a case-by-case basis or steps taken to ensure that tight tolerances on gas quality are adhered to.

In the absence of such data, corrosion resistant alloy (CRA) lines need to be considered.

See also the general recommendations in this RP.

F.5   Corrosion monitoring and data management

The key elements of this RP apply equally to this section.

F.6   Inspection

The key elements of this RP apply equally to this section.

F.7   Review and feedback

The key elements of this RP apply equally to this section.

F.8   Change of duty

The key elements of this RP apply equally to this section.

F.9   Bibliography

- Intergovernmental Panel on Climate Change (IPCC) 2007: Climate Change 2007: Synthesis Report, A Report of the Intergovernmental Panel on Climate Change (IPCC), WMO and UNEP, 2007.
- Intergovernmental Panel on Climate Change (IPCC) 2005: Carbon Dioxide Capture and Storage, Summary for Policymakers and Technical Summary, IPCC Special Report, WMO and UNEP, 2005.

F.10   Reference

1. B. Kermani and F. Daguerre: Proc. CORROSION 2010, 14–18 March 2010, San Antonio, TX. NACE Annual Corrosion Conference, Paper 10334, 2010.

Steam injection is becoming an increasingly common method of extracting heavy oil as part of enhanced oil recovery (EOR). Injecting steam into reservoirs heats oil trapped in rock pores, making it less viscous so it flows more easily. This annex (Fig. G.1) covers and recommends specific corrosion management aspects of steam injection lined within pipeline corrosion management (PCM). It is structured in line with the entire RP and, where appropriate, specifies additional recommendations.

There are several different forms of the technology, with the two main ones being cyclic steam stimulation and steam flooding. Both are most commonly applied to oil reservoirs which are relatively shallow and contain crude oils which are very viscous at the temperatures of the native underground formation. Injecting steam into reservoirs heats oil trapped in rock pores, making it thinner so it flows more easily.

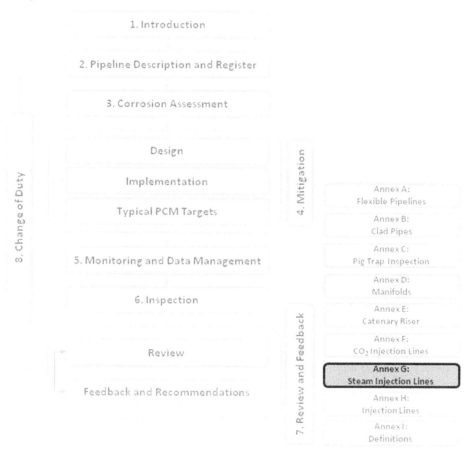

G.1 Position of Annex G within the structure of this RP

G.1   Introduction

An approach to EOR employs high-pressure steam to raise oil production and extend the life of reservoirs. Further details can be found elsewhere outlined in the Bibliography (Section G.9).

G.2   Pipeline description and register

The key elements of this RP apply equally to this section.

G.3   Corrosion assessment

The most common source of corrosion in steam injection systems is the dissolved gases from the feed-water including oxygen, carbon dioxide and ammonia. Amongst these, oxygen is the most aggressive. The importance of eliminating oxygen as a source of pitting and iron deposition cannot be overemphasised. Even small concentrations of this gas can cause serious corrosion problems. Therefore, steam preparation is a key point of consideration.

G.3.1   Relevant corrosion threats

The key elements of this RP apply equally to this section.

Whilst general corrosion is not typically a significant threat for steam injection lines, localised corrosion can be an issue depending on steam temperature and composition.

Typical problems that occur with steam lines are pitting, deposition of insoluble compounds (scale), and corrosion. This damage is mitigated through system design, selection of suitable fabrication materials, process control, and the use of controlled dosages of particular substances that prevent or reduce these problems [1].

In addition, the following corrosion threats should be noted:

- Steam injection lines tend to operate above the temperature range where corrosion under insulation (CUI) becomes a concern. However, lines that operate in cyclic temperature service may be susceptible to CUI and should be designed accordingly.
- Cyclic temperature service could present other issues including internal corrosion from condensing water, thermal fatigue at expansion joints depending on design and welding defects, etc.

G.3.2   Gathering data inputs

In addition to the data outlined in Chapter 3, other parameters that should be compiled include maximum and minimum steam temperatures and any other agents that are present.

G.3.3   Competency

The key elements of this RP apply equally to this section.

G.3.4   Non-routine operations

The key elements of this RP apply equally to this section.

G.3.5   Other hotspots

The key elements of this RP apply equally to this section.

G.3.6   System corrosivity

In the absence of free water, corrosion is not an issue in steam injection lines irrespective of corrosive agents. In contrast, in the presence of free water, the system will experience severe corrosion. In addition, as mentioned earlier, dissolved gases and particularly oxygen exacerbate corrosion. In such cases, similar methods as described in Annex H may be applicable. Most corrosion problems typically occur during upset conditions and especially during down times.

See also the general recommendations in this RP.

G.4   Corrosion mitigation

The key elements of this RP apply equally to this section.

G.4.1   Materials selection

The key elements of this RP apply equally to this section.

In addition, materials selection should be based on standardised materials for elevated temperature application that have minimum defined mechanical properties. For the acceptance of such materials, it should be noted that at least hot yield testing (and possible creep rupture strength testing) may be required to confirm the mechanical properties at elevated temperatures.

Also, appropriate insulation should be selected to minimise heat loss if required.

While linepipe steels according to API 5L-X grades (ISO 3183) are not 'standardised materials for elevated temperature service', they have been used in main steam injection lines. Further information on type of materials is provided in the Bibliography (Section G.9).

G.4.2   Corrosion resistant alloy (CRA) clad and non-metallic liners

The key elements of this RP apply equally to this section.

In addition, non-metallic liners are not considered suitable for steam injection lines due to their temperature limits. However, in the event of upset conditions and especially during down times where water may condense, CRA clad may be considered.

G.4.3   Chemical and physical mitigation methods

The key elements of this RP apply equally to this section.

In addition, the previous sections demonstrate the importance of proper deaeration of feed water to prevent oxygen corrosion. Complete oxygen removal may not be attained by mechanical deaeration alone. Equipment manufacturers state that a properly operated deaerating heater can mechanically reduce the dissolved oxygen concentrations in the feed water to below 0.007 cm^3 L^{-1} (10 ppb) and 0 ppb free carbon dioxide. Typically, steam injection oxygen levels vary from 3 to 50 ppb. Traces of dissolved oxygen remaining in the feed water can potentially be removed chemically with the oxygen scavenger.

G.5   Corrosion monitoring and data management

The key elements of this RP apply equally to this section. In addition, it should be noted that if upset conditions are expected and also during down times where there may be a concern about oxygen ingress and formation of condensed water, precautionary measures similar to those outlined in Annex H should be considered.

G.6   Inspection

The key elements of this RP apply equally to this section.

G.7   Review and feedback

The key elements of this RP apply equally to this section.

G.8   Change of duty

The key elements of this RP apply equally to this section.

G.9   Bibliography

- DIN EN 10216, Seamless Steel Tubes for Pressure Purposes – Technical Delivery Conditions.
- Canadian Standards Association (CSA) Z662, Oil and Gas Pipeline Systems.
- API Spec 5L, Specification for Line Pipe ISO 3183, Petroleum and natural gas industries – Steel pipe for pipeline transportation systems.

G.10   Reference

1. NACE International Resource Centre (www.nace.org).

Water and/or gas injection have proved to be the most economical methods for reservoir management. The technology can be valuable in helping maintain reservoir pressure, enhancing production of hydrocarbon reserves and reducing the environmental impact. This is typically carried out by injection of water or gas. This annex (Fig. H.1) covers and recommends specific corrosion management aspects of water and/or gas injection lines within the pipeline corrosion management (PCM) plan. It is structured in line with the entire RP and, where appropriate, specifies additional recommendations. For gas injection lines, refer to the main body of this RP.

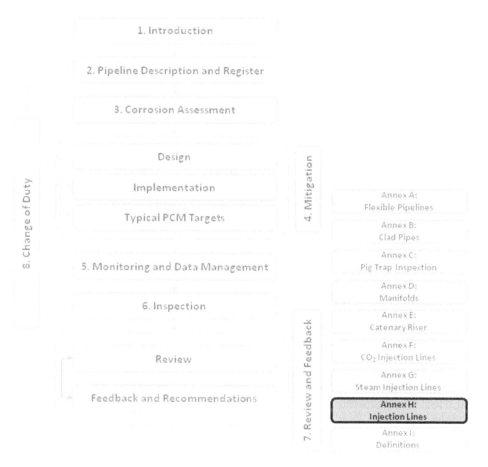

H.1 Position of Annex H within the structure of this RP

H.1   Introduction

Enhanced recovery and increased production are amongst the key challenges facing hydrocarbon production. These entail an increasing trend in the necessity for injection of gas and/or water for reservoir management as well as to minimise environmental impact. The main sources of injected water include:

- treated seawater (SW)
- raw seawater
- aquifer, brackish or river waters
- produced water (PW)
- gas
- comingled waters comprising two or more of these sources but typically a combination of treated SW and PW.

These, their respective corrosion threats and corrosion mitigation methods are outlined in a later section in this annex. Further details can be found elsewhere as outlined in the Bibliography (Section H.9).

H.2   Pipeline description and register

The key elements of this RP apply equally to this section.

H.3   Corrosion assessment

See general recommendations in this RP. Whilst there are established tools for assessing the corrosion threat for single water sources, comingled waters represent a greater challenge and require special care to be taken in assessing potential synergies between the main corrosion mechanisms.

H.3.1   Relevant corrosion threats

Whilst general corrosion is not typically a significant threat for water injection lines, localised corrosion can be an issue depending on water type and chemistry. Particular care should be taken for the case of galvanic corrosion in injection systems (including selection of metal seals). In addition to the threats covered in Chapter 3, specific cases of corrosion threats in water injection lines are summarised herewith and tabulated in Table 4.1.

H.3.1.1   Treated water

Sea, brackish, aquifer or river water treatment before injection involves several stages. These are captured in Table H.1 and usually involve filtration (to remove solids, etc.), treatment with chlorine (as a biocide) and oxygen removal (typical techniques are vacuum deaeration; gas stripping [e.g. using produced gas]; chemical oxygen scavengers [typically bisulphite salts]). In addition, there may be injection of scale inhibitors and periodic injection of chemical biocides. Important parameters include:

i.   **Level of residual dissolved oxygen (ppb by weight).** This will include the typical background level of oxygen (typical target range is less than 30 ppb); details of any 'spikes' in the dissolved oxygen levels, e.g. due to 'trips' in the oxygen

Table H.1 Summary of corrosion threat and evaluation methods in injection lines

Media	Corrosion threat	Manifestation	Principal parameters	Verification/standard/guideline	Mitigation
Produced water (PW)	Primarily CO_2/H_2S corrosion, MIC (possible O_2)	Metal loss (general and/or localised corrosion/cracking)	• CO_2 • H_2S • Biocide • Operating conditions • Water chemistry (possible O_2)	Many models, no international standard (H_2S cracking according to ISO 15156)	Acid gas removal, drying or inhibition & biocide treatment
Sea, brackish, aquifer, or river waters	Primarily O_2 corrosion, MIC	Metal loss (general and/or localised corrosion)	• O_2 • Cl_2 • Biocide • Operating conditions • Water chemistry	Many models, no international standard	O_2 removal & biocide treatment
SW/PW (comingled water)	Both O_2/Cl_2 and CO_2/H_2S; MIC, scale solids	Metal loss (general and/or localised corrosion)	• CO_2 • H_2S • O_2 • Biocide • Operating conditions • Water chemistry	No proven model, no international standard	Acid gas and O_2 removal & biocide treatment, scale inhibition
Gas injection	Primarily CO_2/H_2S corrosion, organic acids	Metal loss (general and/or localised corrosion/cracking)	• CO_2 • H_2S • Operating conditions • Water chemistry	Many models, no international standard (H_2S cracking according to ISO 15156)	Drying or inhibition & biocide treatment
Water alternating gas (WAG)	Both O_2 and CO_2/H_2S	Metal loss (general and/or localised corrosion)	• CO_2 • H_2S • O_2 • Biocide • Operating conditions • Water chemistry	No proven model, no international standard	Acid gas and O_2 removal & biocide treatment

removal equipment, during chemical biocide injection (number, magnitude [ppb], length of time [minutes or hours]).

ii. **Presence and levels of any dissolved acidic gases (CO_2, H_2S, organic acids).** These can be introduced via the produced gas if gas stripping is used or in comingled (treated seawater and produced water) water injection systems.

iii. **Residual chlorine levels (ppm weight, wt.).** Chlorine is normally injected before oxygen removal, but it is usual to have residual chlorine in the water after oxygen removal. Some operating companies aim to completely remove chlorine before injection whilst others aim for a maximum level of 0.2 or 0.3 ppm wt., depending on the water quality requirements (see section H.4.4).

H.3.1.2   Raw seawater

As the name implies, the seawater in this case receives no treatment apart from filtration to remove solids, the addition of chlorine, to act as a biocide and nitrate to control sulphate-reducing bacteria (SRB) growth. Important parameters are:

- levels of chlorine (ppm weight)
- level of solids, i.e. sand (sand size [microns]; sand concentration [lbs per thousand barrels of water])
- level of nitrate.

H.3.1.3   Produced water

It is becoming increasingly common to inject produced water as a means of water disposal. The produced water is taken from the separators and usually goes to a degassing drum, holding tank or similar to allow as much of the corrosive dissolved gases as possible (H_2S, CO_2) to dissipate before injecting the water. The produced water corrosivity is established by considering the conditions (partial pressure of corrosive gases) in the last stage of equilibration between water and gas (e.g. water storage vessel). Important parameters include:

i. **CO_2/H_2S level:** These should represent the levels of dissolved gases in the injected water. It is normal to take this as the partial pressure of the gases in the last stage where the water and gas equilibrate (e.g. typically in the holding tank or, if there is no such tank, in the last degassing drum before injection).

ii. **Produced water chemistry**: This should include full water analysis (including Na^+, Ca^{2+}, K^+, Mg^{2+}, Fe^{2+}, Ba^{2+}, Sr^{2+}, Cl^-, S^{2-}, SO_4^{2-}, HCO_3^-], amounts of organic salts/acids (including acetate, propionate, butyrate) and pH.

iii. **Presence and levels of nitrates**: These can be intentionally added, typically as calcium nitrate, as a means of reducing the probability of reservoir souring and can adversely affect the injection water corrosivity.

iv. **Presence of corrosion inhibitor:** The produced water may well contain residual corrosion inhibitor from treating the fluids in the production stream (or even corrosion inhibitor purposely added to the produced water stream) that may afford a degree of protection from corrosion.

v. **Presence of scale inhibitor:** The produced water may contain residual scale inhibitor, the presence of which can affect corrosivity of the injected water.

vi. **Presence of biocide:** This can be intentionally added continuously or in batch dose to control bacteria proliferation. Some biocides may be corrosive.

vii. **Dissolved oxygen levels:** Dissolved oxygen does not naturally occur in produced waters, but may be present in small amounts due to low pressure equipment, air from leakage at the pump seals, or stand-by streams. This also adversely affects system corrosivity.

H.3.1.4   Comingled water

Where there is inadequate produced water alone for reservoir maintenance, it is necessary to supplement the injection produced water. Treated seawater is normally used for this function. This mixture of produced water and treated seawater is often known as 'comingled water'. Important parameters include:

i.   **Levels of dissolved corrosive gases (O_2/H_2S/CO_2):** This should be either in the form of the equivalent partial pressure of gas (for H_2S/CO_2 originating from the produced water) or ppm by wt. (e.g. levels of dissolved O_2 for water from the seawater treatment plant). In the case of dissolved oxygen levels this should include the typical background level of oxygen (it is normal to aim for a level of 10 ppb or less); details of any 'spikes' in the dissolved oxygen levels, e.g. due to 'trips' in the oxygen removal equipment, during chemical biocide injection (number, magnitude [ppb], length of time [minutes or hours]).

ii. **Produced water chemistry:** This should include full water analysis including Na^+, Ca^{2+}, K^+, Mg^{2+}, Fe^{2+}, Ba^{2+}, Sr^{2+}, Cl^-, S^{2-}, SO_4^{2-}, HCO_3^-], amounts of organic salts/acids (including acetate, propionate, butyrate) and pH.

iii. **Treated seawater chemistry:** This should include the levels of residual chlorine and any other water treatment chemicals added, e.g. biocides, oxygen scavenger.

iv. **Presence of corrosion inhibitor:** The produced water may well contain residual corrosion inhibitor from treating the fluids in the production stream (or even corrosion inhibitor purposely added to the produced water stream). While the inhibitor may afford a degree of protection, the performance of the type of inhibitors used can be adversely affected by the presence of dissolved oxygen (resulting from the comingled seawater). Under such circumstances, care needs to be exercised to ensure that the target O_2 residual levels (typically either 10 ppb or 20 ppb) are being consistently met, since extended exposure to higher levels of dissolved oxygen in the presence of corrosion inhibitor may promote pitting corrosion. Corrosion inhibitor packages containing quaternary ammonium/amine salts (namely, these also have biocidal properties) offer the most tolerance to dissolved oxygen.

In addition, consideration should be given to those outlined in Section H.3.1.3.

H.3.1.5   Gas

Typically gas for injection will be 'dried' (i.e. water removed such that the gas is above the water dew point temperature) before injection. Therefore, in principle, there should be no corrosion issues. However, even dried gas will contain some water (in the gaseous form) such that some checks are required to ensure the gas will remain dry (i.e. there will be no free/condensing water) in service under all circumstances. In addition, there have been discussions about the possibility of using wet gas for injection in the future. The injection of wet gas is outside the scope of this annex but generally, such pipelines should be treated in much the same way as a wet gas production well. Important parameters include:

- water dew point temperature and pressure for the injection gas (this will help determine whether the downhole completion will be 'water wetted')
- partial pressure of H_2S and CO_2 in the injection gas.

H.3.1.6   Water alternating gas (WAG)

The WAG injection process aims to enhance the oil recovery process. In WAG, water and gas are injected alternately for periods of time. This process is used mostly in CO_2 floods to improve hydrocarbon contact time and sweep efficiency of the CO_2. The corrosion process in WAG would be subject to gas and water types and corrosion threat would be associated with both oxygen and CO_2/H_2S corrosion as outlined in Table H.1.

H.3.1.7   General concerns

In addition to the above, solids will need to be considered in the same way as for production pipelines and should be managed accordingly.

In addition to Table H.1, two specific types of threat in water injection lines are:

i.   groove-line corrosion at 6 o'clock locations
ii. preferential weld corrosion.

Note that the means of mitigating the main corrosion threats can also create further threats. For example, hypochlorite injection or electro-chlorination is commonly employed in the early stages of seawater treatment to inhibit bacterial activity, but residual Cl_2 must be controlled to prevent accelerated corrosion in the pipework. Similarly, excessive residual oxygen scavenger can prove aggressive, particularly at elevated temperatures. Careful control is needed in injection systems to balance the mitigation of the various threats. Such care is required for both C-steel and corrosion resistant alloy (CRA) pipelines.

H.3.2   Gathering data inputs

The key elements of this RP apply equally to this section.

H.3.3   Competency level requirements

The key elements of this RP apply equally to this section.

H.3.4   Non-routine operations

The key elements of this RP apply equally to this section.

H.3.5   Other hotspots

The key elements of this RP apply equally to this section.

H.3.6   System corrosivity

The principal types of corrosion in water injection lines and methods of assessment are included in Table H.1.

It has been recognised that high oxygen, short-term excursions during water injection are likely. Even in fully functional oxygen removal systems, injection of oxygen scavenger is typically halted during biocide injection resulting in oxygen spikes. System availability should be considered in the assessment of water corrosivity.

It should be noted that dissolved oxygen can be fairly quickly consumed by C-steel pipelines. Therefore, the greatest threat of oxygen corrosion is typically in the first kilometre or so of the line. The long length of piping between deaeration and pipeline entry may reduce potential corrosivity. However, replacing these areas with CRA or non-metallic material would transfer oxygen downstream.

See also general recommendations in this RP.

H.4   Corrosion mitigation

In addition to those outlined in Chapter 4, mitigation systems in water injection lines include (i) water treatment and (ii) materials upgrade. Some aspects of these are coved in Table H.1, and with further details throughout the RP.

H.4.1   Materials selection

Materials options for water injection lines C-steel and corrosion-resistant options are given in Annex B.

H.4.2   CRA clad and non-metallic liners

The key elements of this RP apply equally to this section. However, non-metallic liners which may be considered for water injection lines are covered in Annex B.

H.4.3   Chemical and physical mitigation methods

The key elements of this RP apply equally to this section. In addition, specific recommendations are included in Table H.1.

H.4.4   Typical PCM targets for water injection lines

Water quality is paramount in affording low corrosion of C-steel pipelines. Typical PCM monitoring parameters for water injection lines are summarised in Table H.2. The required water quality is dependent on many factors including inspectability and the consequences of leakage. The parameters shown in Table H.2 are for stringent criteria and applicable to critical and uninhibited lines.

H.5   Corrosion monitoring and data management

See general recommendations in this RP. In addition, monitoring of bacterial activity and oxygen monitoring are more important than other applications.

H.6   Inspection

The key elements of this RP apply equally to this section.

Table H.2 Typical PCM monitoring parameters for water injection lines

Parameter	Typical PCM target for water injection lines
Oxygen content	Removal of residual oxygen content typically maintaining to below 20 ppb (excursions to higher values are typical, for example, when injecting biocide)
Solids content	Water should have low solids content, for example, solids < 1 lb/1000 bbl

Residual Cl_2 or biocide content	• Water should contain low residual chlorine (for example < 0.3 ppm) in the system – an organic biocide (such as glutaraldehyde) treatment is typical choice when using CRAs
	• A stringent biocide programme from the beginning of the water handling programme is strongly recommended
	• Examples of sessile bacteria limit:
	  o SRB < 10^2 cm^{-2}
	  o GAB, GnAB < 10^2 cm^{-2}
	• Examples of planktonic bacteria
	  o SRB < 1 ml^{-1}
	  o GAB, GnAB < 10^4 ml^{-1}

Velocity	Water velocity should be maintained below operational limits to control any fluid erosion threats; for example, API RP 14E limits

H.7   Review and feedback

The key elements of this RP apply equally to this section.

H.8   Change of duty

The key elements of this RP apply equally to this section.

H.9   Bibliography

- J. R. Postgate: 'The sulphate-reducing bacteria', 2nd edn, 43; 1984, Cambridge, Cambridge University Press.
- C. A. C. Sequeira, ed.: 'Microbial corrosion', EFC Publication 29; 2000, London, The Institute of Materials.
- J. G. Stoeckler II, ed.: 'A practical manual on microbiologically influenced corrosion', Vol. 2; 2001, Houston, TX, NACE International.
- 'Microbiologically influenced corrosion and biofouling in oilfield equipment', TCP 3; 1990, Houston, TX, NACE International.
- S. W. Borenstein: 'Microbiologically influenced corrosion handbook'; 1994, New York, Industrial Press.

A limited number of definitions are described in this annex within the context of the present RP. They are neither exhaustive nor systematically accepted throughout the petroleum industry. Detailed corrosion related definitions and terminologies can be found elsewhere [1].

Asset Integrity Management (AIM):	AIM refers to the management of systems, strategies and activities aimed at maintaining the plant in fit-for-purpose condition for its desired life. It involves development and implementation of a series of continual measures to enable safe, green, secure and continued production throughout the planned design life ensuring availability, reliability and integrity. AIM incorporates aspects of design, operations, maintenance, and inspection.
Corrosion Management:	Corrosion management is that part of the overall AIM, which is concerned with the development, implementation, review and maintenance of the corrosion policy [2].
Data Management System:	A system developed to acquire, store and manage data preferably in an auditable manner.
Fluids:	In the context of this RP, fluids include gas, water and hydrocarbon or a mixture.
In-Line Inspection (ILI):	An inspection of a pipeline from the interior of the pipe using an in-line inspection tool. Also called intelligent or smart pigging [3].
In-Line Inspection Tool (ILI Tool):	The device or vehicle that uses a non-destructive testing (NDT) technique to inspect the pipeline from the inside. Also known as intelligent or smart pig [2].
Pig:	A generic term signifying any independent, self-contained or tethered device, tool, or vehicle that moves through the interior of the pipeline for inspecting, dimensioning, or cleaning. A pig may or may not be an in-line inspection tool [2].
Pipeline:	A line which carries fluids from wellheads (including flowlines) or production sites (including trunklines) to process sites (including platforms, reception plants, and downstream facilities) and vice versa, excluding wells, vessels, rotating equipment and process pipework.
Pipeline Corrosion Management (PCM):	PCM is defined as that part of the overall pipeline integrity management system which is concerned with the development, implementation, review and maintenance of the corrosion policy.

I.1   References

1. 'Standard terminology and acronyms relating to corrosion', NACE/ASTM G193-09, NACE International/ASTM International, 2009.
2. NACE International (www.nace.ord), USA.
3. 'In-line inspection of pipelines', NACE Standard RP0102-2002, NACE International, Houston, TX, 2002.